Petroleum Geoscience

Petroleum Geoscience

Editor

Raju Apte

Petroleum Geoscience

Edited by **Raju Apte**

Printed in 2017

ISBN: 978-1-68117-150-0

Library of Congress Control Number: 2015936565

© 2016 by
SCITUS Academics LLC,
616, Corporate Way, Suite 2, 4766,
Valley Cottage, NY 10989

www.scitusacademics.com

Contents

Preface

Petroleum Geoscience is a refereed journal providing a multidisciplinary outlet for the publication needs of those involved in the science and technology associated with rock-related petroleum disciplines. The journal has a circulation of approximately 5000 and includes in its readership geologists, geophysicists, petroleum engineers, petro physicists and geochemists in the academic and professional worlds. The journal aims to improve knowledge of petroleum geoscience and reflect the international nature of the petroleum industry. Articles range from exploration geology, geophysics, geochemistry and petro physics to reservoir geology and reservoir engineering. Petroleum Geoscience highlights the optimization of resources and maximization of quality and profit through the application of technology and brings the benefits of the latest developments to its wide readership.

Editor

Chapter 1

Seismic Expression of Polygonal Faults and its Impact on Fluid Flow Migration for Gas Hydrates Formation in Deep Water of the South China Sea

Duanxin Chen[1,2], Shiguo Wu[1], Xiujuan Wang[1,] and Fuliang Lv[3]

[1]Key Laboratory of Marine Geology and Environment, Institute of Oceanology, Chinese Academy of Sciences, Qingdao 266071, China

[2]Graduate University of Chinese Academy of Sciences, Beijing 100049, China

[3]Hangzhou Institute of Geology, PetroChina, Hangzhou 310023, China

ABSTRACT

Polygonal faults were identified from three-dimensional (3D) seismic data in the middle-late Miocene marine sequences of the South China Sea. Polygonal faults in the study area are normal faults with fault lengths ranging from 100 to 1500 m, fault spaces ranging from 40 to 800 m, and throws ranging from 10 to 40 m. Gas hydrate was inferred from the seismic polarity, the reflection strength, and the temperature-pressure equilibrium computation results. Gas hydrates located in the sediments above the polygonal faults layer. Polygonal faults can act as pathways for the migration of fluid flow, which can supply hydrocarbons for the formation of gas hydrates.

INTRODUCTION

Polygonal faults are a network of layer-bound, mesoscale (throws from 10 to 100 m) extensional faults arranged in a polygonal structure developing in deep-water sequence [1]. The term "polygonal fault" was named by Cartwright [2] when he analyzed the shale sedimentary throughout the 3D seismic data obtained from North Sea basin. Up to now, more than 50 basins have found the existence of polygonal faults. Some geologists presented different formation mechanisms, including density inversion [3, 4], gravity sliding or collapse [5], episodic hydrofracturing [6], "volumetric contraction" [7], and low coefficients of friction [8].

In the South China Sea (SCS), polygonal faults were for the first time identified in the Qiongdongnan basin (QDNB) in 2009 [9]. Sun et al. [10] mentioned the migration of hydrocarbon through polygonal faults in QDNB. However, there is no document about reservoirs of gas hydrates associated with polygonal faults in SCS.

In this paper, we will show the seismic geometry and distribution of polygonal faults and investigate the geophysical characters of gas hydrates in deep water of QDNB and Zhongjiannan basin (ZJNB) (Figure 1). We use high-resolution 2D and 3D seismic data integrated with local sedimentary history to study the function of polygonal faults in acting as conduits of gas hydrate provinces.

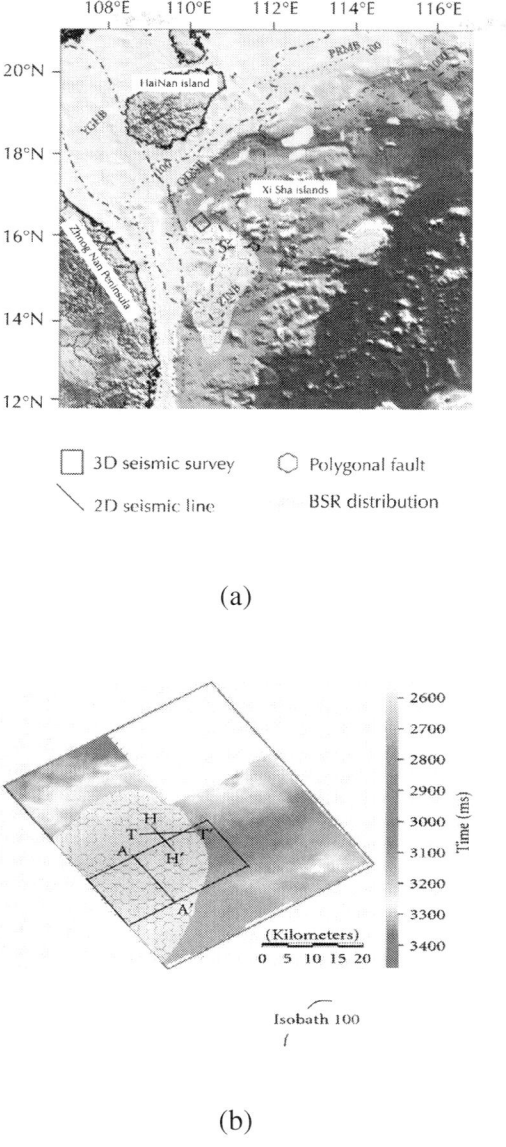

(a)

(b)

Figure 1: (a) The schematic map of South China Sea. The map shows the distribution of polygonal faults and prospective gas hydrate zones in QDNB and ZJNB. (b) The 3D seismic survey in deep water of QDNB. The basemap is the time depth of the sequence boundary of the base of upper Miocene. Lines plotted in the map will be discussed below.

GEOLOGICAL BACKGROUNDS

Sedimentary basins in northern South China Sea margin underwent an early syn-rifting stage in Paleogene and a postrifting thermal subsidence stage in Neogene and Quaternary [11]. The separation for syn-rifting and postrifting stage is a breakup unconformity T60 (stratigraphic frames listed in Figure 2) between Oligocene and Miocene. Sedimentary facies in syn-rifting stage chiefly evolved from alluvial, delta, and lacustrine facies to shallow marine facies, whereas shallow, bathyal, and abyssal facies occupied the postrifting stage [11]. Sediments in lower Huangliu formation and upper Meishan formation of Miocene sequence consist of fine-grained claystone and limestone-hosting polygonal faults based on wireline log interpretations from wells drilled in QDNB.

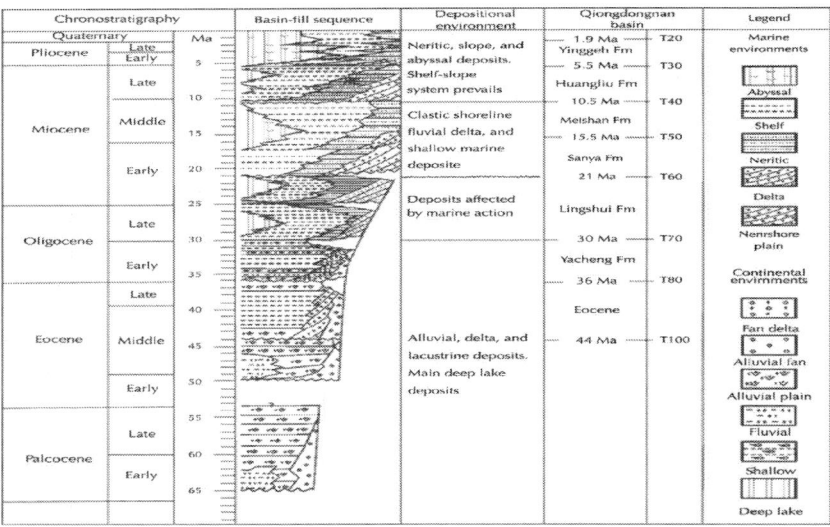

Figure 2: Chronostratigraphy chart illustrating the basin-fill sequence and depositional environment of QDNB and ZJNB, modified from Xie et al. [11].

DATA AND METHODOLOGY

This study is based on an integrated analysis of seismic data including a 3D survey and 2D lines acquired by BGP Inc., (China National

Petroleum Corporation (CNPC)), processed by Liaohe Petroleum (CNPC, PetroChina). The 3D seismic data covers an area of 1480 km² with a bin size of 12.5m × 25m and water depth ranging from 1120 m to 1275 m in west of Xisha Islands, while 2D seismic data covers the deepwater of the southern depression of QDNB and the overall ZJNB with the common depth point (CDP) interval of 25 m (Figure 1(a)). We used the coherency slice and sequence flattening technique to study the 3D and partial 2D seismic data. With estimated geothermal gradient values and proper formation velocities, the study utilizes temperature-pressure equilibrium to compute the base of GHSZ.

RESULTS

Seismic Imaging to Identify Polygonal Faults

Polygonal faults were identified in deep water of the southern depression of QDNB and northeastern of ZJNB over an area of about 6000 km² (Figure 1(a)). 3D seismic data from QDNB indicated that the polygonal faults pervasively occur in the lower Huangliu formation and upper Meishan formation and are limited by layer-bound strata of the base of Pliocene (T30) and Meishan formation (T50), as seen in Figure 3. Polygonal faults in northern SCS are extensional faults with fault lengths ranging from 100 to 1500 m, fault spaces ranging from 40 to 800 m, and throws ranging from 10 to 40 m (Figures 3 and 4). The view of the coherence slice from the 3D seismic data shows the irregular polygonal map geometry (Figure 4). The deformed interval is generally subdivided into two tiers. The upper tier lies in the layer of 2.5–2.7 s two-way travel time (TWT)- and the lower one is in the layer of 2.7–3.0 s (TWT), manifesting the stratigraphically delimited subunits within the deformed interval. However, faults with longer vertical slip may connect the two tiers, piercing the base of Pliocene and Meishan formation. Smaller faults could be found dim among distinct polygonal faults (Figure 4). It might indicate the existence of smaller scale of polygonal faults.

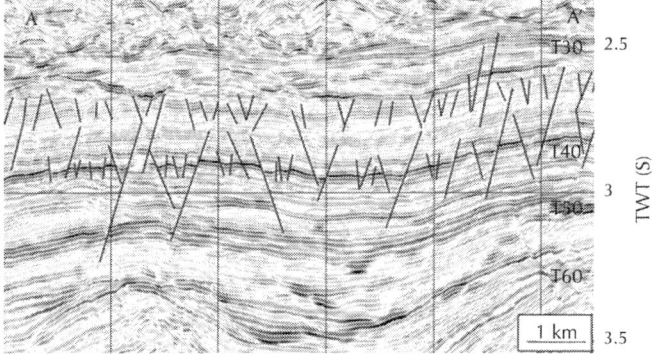

Figure 3: Seismic profile of line AA' showing vertical characteristics of polygonal faults, for example normal faults with throws ranging from 10 to 40 m, fault spaces from 40 to 800 m, and two tiers visible.

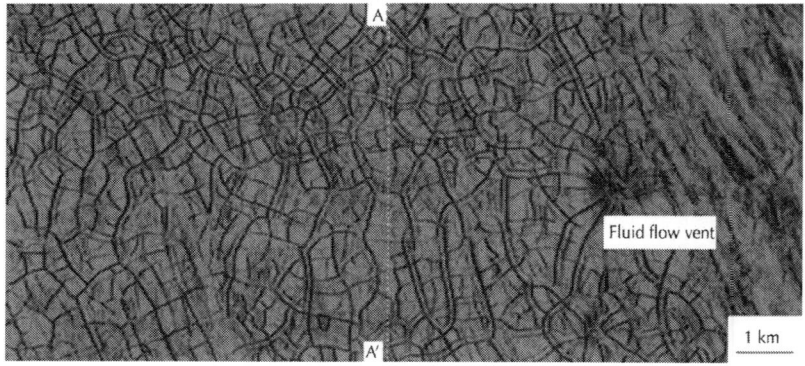

Figure 4: Coherence slice through the T40 horizon. The polygonal geometry is visible in the map. The rectangle boundary and vertical seismic profile of line AA' are displayed in Figure 1(b). The vent may have formed as a result of expelled fluid flow.

BSRs Occurrence in Deep Water of SCS

Distributions and characteristics of BSRs in deep water of QDNB have been delineated (Figure 1(a)) [12–16]. Most of the BSRs with middle-to- high amplitude develop in Neogene and Quaternary strata which are paralleling to the seafloor.

We calculated the base of GHSZ and plotted it in seismic profile SS' in Figure 5 using the geothermal gradient and water depth obtained from this basin. The geothermal gradient was 0.045°Cm⁻¹ suggested by Ma et al. [17]. The water depth was acquired in accordance with multibeam data.

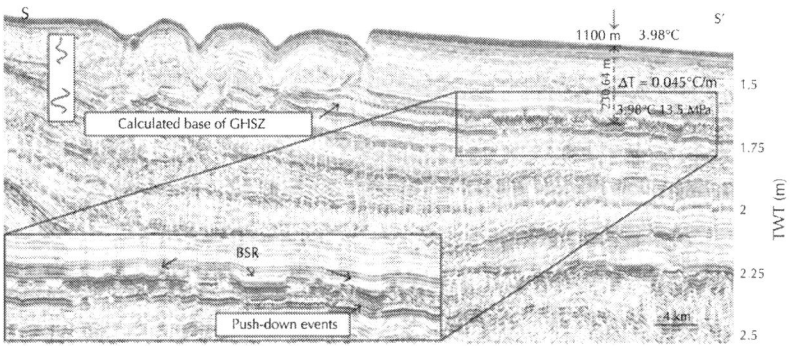

Figure 5: The calculated depth of GHSZ, and heat flow values in seismic profile SS'.

Temperature-pressure equilibrium equation was given by Miles [18]

$$P = 2.8074023 + a \cdot T + b \cdot T^2 + c \cdot T^3 + d \cdot T^4, \tag{1}$$

where $P(MP_a)$ is the formation pressure; T (°C) is formation temperature; $a = 1.559574 \times 10^{-1}$; $b = 4.8275 \times 10^{-2}$; $c = -2.78083 \times 10^{-3}$; $d = 1.5922 \times 10^{-4}$. And values of parameters (ad) were used by Wang et al. [19].

In (1), the input water depth is 1100 m and the geothermal gradient is 0.045°C m⁻¹. The thickness of base of GHSZ is 230.64 m and the pressure of the BSR is about 13.53 MPa. The conductivity is 1.18792 Wm⁻¹°C⁻¹, and heat flow value at the BSR is 53.456 mWm⁻². The result is plotted in Figure5. Heat flow value is in accordance with He et al. [20], who proposed that the average heat flow was about 59 mWm⁻² which was caused by the high rate of sedimentation. The estimated thickness of GHSZ also coincides with the values given by Jin et al. [21] and Shi et al. [22] in this area.

DISCUSSIONS

Focused Fluid Flow Associated with Polygonal Fault

Many people have reported the existence of pockmarks, gas chimneys, mud diapirs, slides, and channels and discussed their relationship to the fluid flow in northern South China Sea [14, 16, 19, 23, 24]. However, the polygonal faults may provide an additional pathway for the fluid flow, and the relationship has also been documented in the Lower Congo Basin [25], Norway continental margin [26, 27], and Scotian slope of the eastern Canada continental margin [28].

In QDNB and ZJNB, a large volume of source rocks and syn-rift faults were well developed in the Paleogene sequences, whereas few faults occurred in the postrifting stage in deep water, which limited deep hydrocarbon in syn-rifting strata to migrate upward into the post-rifting strata. Therefore, polygonal faults developing in the lower Huangliu formation and the upper Meishan formation provide enhanced permeability for the fine-grained claystone and act as a pathway linking source rock in the syn-rifting strata to the reservoir in the post-rifting strata which may release more hydrocarbons migrating upwards the base of GHSZ.

Fluid flow involved in mud diaper, upwelling along a polygonal fault, indicates that polygonal faults may serve as fluid flow conduits, as shown in Figure 6. In the section, fluid flow commenced its migration through the lower tier of polygonal faults. Acoustic blanking zones and strong reflectors above sacking structures indicate the presence of gas. A small amount of fluid could also flow upwards via the upper tier. Polygonal faults display a radiant shape, as shown in Figure 4.

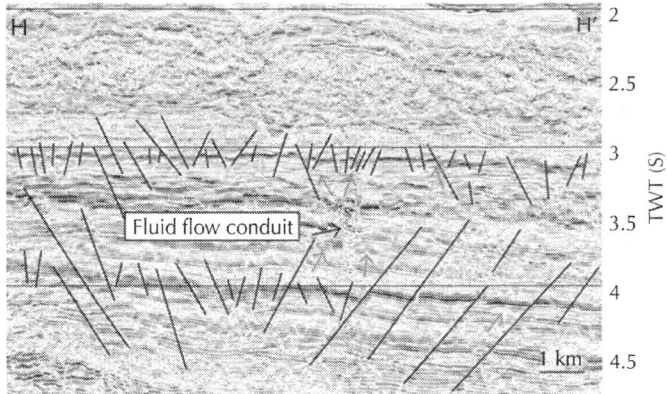

Figure 6: Interpretation of seismic profile HH', which describes the migration of fluid flow through polygonal faults. The dim seismic reflectors and fluid flow pipes are visible. See position in Figure 1(b).

The model of low coefficient of friction raised by Goulty [8] may support the function of conduits of polygonal faults as well as conventional faults. It was demonstrated that low coefficients of residual friction in fine-grained sediments might be key to the formation of the polygonal system, and there was no evidence that nucleation of those faults that evolve into polygonal systems differs fundamentally from the processes involved in the nucleation of conventional faults in soft sediments [8]. Previous studies show that polygonal faults often initiate at shallow burial depth [2,7, 29]. With increasing buried depth, compacted dewatering continuously occurs in layers underlying polygonal faults, and expelled fluids migrate up through polygonal faults and rock porosities to arrive at the seabed. However, with relation to different formation models of polygonal faults, the transporting mechanisms of fluid flows through polygonal faults are controversial [6, 8]. The preferred way of fluid flows may be episode eruptions provided by Cartwright [6]. Roberts and Nunn (1995) also conducted numerical simulations showing that a vertical fracture opened for 20–50 years in the seal and provided a permeable pathway from the geopressured sediments into the overlying section when the fluid pressure exceeds 85% of the lithostatic pressure. Then Fluid pressure in the seal decreased, and the fracture closed until fluid pressure increased to the fracture criteria, which took 10.000–500.000 years [30].

Gas Hydrate Occurrence with Polygonal Faults

The fact that deep fluid flows upwell through polygonal faults to produce gas hydrates is confirmed in several basins. The relationship between the polygonal fault and gas hydrate has been discussed on the continental margin of Norway, the Lower Congo basin, and the Scotian margin of the east coast of Canada [23, 24, 26]. On the mid-Norwegian margin, the polygonal fault system is located in the fine-grained hemi-pelagic sediments of the Kai formation, showing that fluid flow related to polygonal fault in this area is an ongoing process since the early Miocene in the mid-Norwegian [26,27].

Besides tectonic faults, diapirs, and gas chimney act as conduits for the reservoir of gas hydrates in the South China Sea; the polygonal fault system is considered to be a new pathway linking to the GHSZ. The BSRs locate at the base of GHSZ which may indicate the presence of gas hydrates. The distribution of BSRs is shown in Figure 1(a). The BSRs are discontinuous and occur overlying the polygonal faults. The negative relieves are discovered below the BSRs, which may result from pushing down seismic events due to high concentration of free gas below GHSZ (Figures 5 and 7).

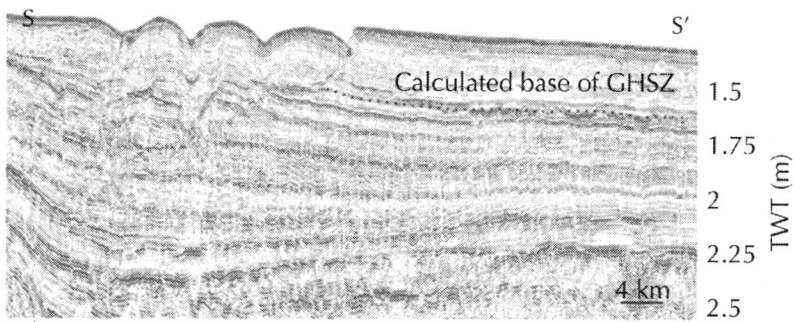

(a)

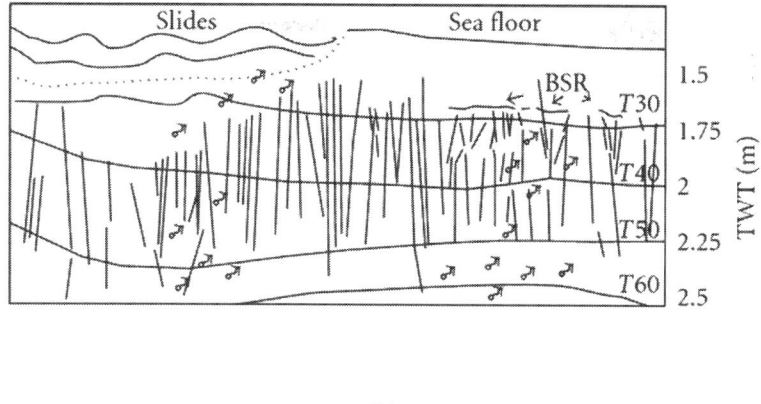

(b)

Figure 7: (a) The seismic profile SS′ (b) The interpretation of line SS′ The figure shows fluid flow migrated into the GHSZ through the polygonal faults. BSRs are distinctly visible overlying the polygonal faults. See the position in Figure 1(a).

Biogenic gas migration may occupy partial gas component in gas hydrate in and above middle Miocene, while thermogenic hydrocarbon source rocks host in early Oligocene Yacheng formation, late Oligocene Lingshui formation, and Miocene Sanya formation [31]. Zhu et al. [32] tested gases released from acid hydrocarbon on 127 shallow sediment samples in QDNB, and indicated the existence of thermogenic gases with methane concentration $10 \sim 243.5\,\mu Lkg^{-1}$, carbon isotope $-43.8\% \sim -26.6\%$, and $C1/(C2+C3)10 \sim 30$ It is inferred that focus fluid flows with large volume of thermogenic gas from the source rocks, together with shallow biogenic gas, may mix and migrate upward along the polygonal fault system and form gas hydrates in locations of appropriate temperature and pressure. Hence, polygonal faults serve as a pathway linking thermogenic gas and shallow biogenic gas to the reservoirs of gas hydrates.

CONCLUSIONS

Researches based on high resolution 3D and 2D seismic data show that polygonal faults are tensional normal faults with fault lengths from 100 to 1500 m, fault spaces from 40 to 800 m, and throws from 10 to 40 m.

The layer-bound sequence surfaces, two or more tiers, and polygonal geometry view pattern in plane can be recognized. Occurrence of gas hydrates in deep water of South China Sea is considered to be related to fluid flow through the polygonal fault system. Polygonal faults link thermogenic gas and shallow biogenic gas to GHSZ, serving as pathways to migrate gas upwards to form gas hydrates.

ACKNOWLEDGMENTS

The authors thank the Hangzhou Institute of Geology of PetroChina for permission to release these data in this paper. They gratefully acknowledge the reviewers for further suggestions and grammar revision. The research work was supported by the National Natural Science Foundation of China (no. 40930845), International Science & Technology Cooperation program of China (no. 2010DFA21740), and the National Basic Research Program (no. 2009CB219505).

REFERENCES

1.   J. A. Cartwright and D. N. Dewhurst, "Layer-bound compaction faults in fine-grained sediments," Bulletin of the Geological Society of America, vol. 110, no. 10, pp. 1242–1257, 1998.

2.   J. A. Cartwright, "Episodic basin-wide hydrofracturing of overpressured Early Cenozoic mudrock sequences in the North Sea Basin," Marine and Petroleum Geology, vol. 11, no. 5, pp. 587–607, 1994.

3.   J. P. Henriet, M. de Batist, and W. van Vaerenbergh, "Seismic facies and clay tectonic features of the Ypresian clay in the Southern North Sea," Bulletin van de Belgische Vereniging voor Geologie, vol. 97, pp. 457–472, 1989.

4.   J. P. Henriet, D. Batist, and M. Verschuren, "Early fracturing of Palaeogene clays, southernmost North Sea: relevance to mechanisms of primary hydrocarbon migration," inGeneration, Accumulation and Production of Europe's Hydrocarbons, A. M. Spencer, Ed., pp. 217–227, Oxford University Press, Oxford, 1991.

5. W. G. Higgs and K. R. McClay, "Analogue sandbox modeling of Miocene extensional faulting in the Outer Moray Firth," Geology Society of London, vol. 71, pp. 141–162, 1993.

6. J. A. Cartwright, "Episodic basin-wide fluid expulsion from geopressured shale sequences in the North Sea Basin," Geology, vol. 22, no. 5, pp. 447–450, 1994.

7. J. A. Cartwright and L. Lonergan, "Volumetric contraction during the compaction of mudrocks: a mechanism for the development of regional-scale polygonal fault systems,"Basin Research, vol. 8, no. 2, pp. 183–193, 1996.

8. N. R. Goulty, "Geomechanics of polygonal fault systems: a review," Petroleum Geoscience, vol. 14, no. 4, pp. 389–397, 2008.

9. S. Wu, Q. Sun, T. Wu, S. Yuan, Y. Ma, and G. Yao, "Polygonal fault and oil-gas accumulation in deep-water area of Qiongdongnan Basin," Shiyou Xuebao/Acta Petrolei Sinica, vol. 30, no. 1, pp. 22–26, 2009 (Chinese).

10. Q. Sun, S. Wu, F. Lü, and S. Yuan, "Polygonal faults and their implications for hydrocarbon reservoirs in the southern Qiongdongnan Basin, South China Sea," Journal of Asian Earth Sciences, vol. 39, no. 5, pp. 470–479, 2010.

11. X. N. Xie, R. D. Müller, S. Li, Z. Gong, and B. Steinberger, "Origin of anomalous subsidence along the Northern South China Sea margin and its relationship to dynamic topography,"Marine and Petroleum Geology, vol. 23, no. 7, pp. 745–765, 2006.

12. B. C. Yao, "Gas hydrates in the South China Sea," Journal of Tropical Oceanography, vol. 20, no. 2, pp. 20–28, 2001 (Chinese).

13. B. C. Yao, "The forming condition and distribution characteristics of the gas hydrate in the South China Sea," Marine Geology & Quaternary Geology, vol. 25, no. 2, pp. 81–90, 2005 (Chinese).

14. D. F. Chen, X. X. Li, and B. Xia, "Distribution of gas hydrate stable zones and resource prediction in the Qiongdongnan basin of the South China Sea," Chinese Journal of Geophysics, vol. 47, no. 3, pp. 483–489, 2004.

15. S. Wu, G. Zhang, Y. Huang, J. Liang, and H. K. Wong, "Gas hydrate occurrence on the continental slope of the northern South China Sea," Marine and Petroleum Geology, vol. 22, no. 3, pp. 403–412, 2005.

16.   X. J. Wang, S. G. Wu, D. D. Dong, Y. H. Gong, and C. Chai, "Characteristics of gas chimney and its relationship to gas hydrate in Qiongdongnan basin," Marine Geology & Quaternary Geology, vol. 28, no. 3, pp. 103–108, 2008 (Chinese).

17.   Q. Ma, S. Chen, Q. Zhang, S. Guo, and S. Wang, Overpressure Basins and Hydrocarbon Distribution, Geological Press, Beijing, China, 2000.

18.   P. R. Miles, "Potential distribution of methane hydrate beneath the European continental margins," Geophysical Research Letters, vol. 22, no. 23, pp. 3179–3182, 1995.

19.   X. Wang, S. Wu, S. Yuan et al., "Geophysical signatures associated with fluid flow and gas hydrate occurrence in a tectonically quiescent sequence, Qiongdongnan Basin, South China Sea," Geofluids, vol. 10, no. 3, pp. 351–368, 2010.

20.   L. He, K. Wang, L. Xiong, and J. Wang, "Heat flow and thermal history of the South China Sea," Physics of the Earth and Planetary Interiors, vol. 126, no. 3-4, pp. 211–220, 2001.

21.   C. S. Jin, J. Y. Wang, and G. X. Zhang, "Factors affecting natural gas hydrate stability zone in the South China Sea," Mineral Deposits, vol. 24, no. 5, pp. 388–397, 2005 (Chinese).

22.   X. Shi, X. Qiu, K. Xia, and D. Zhou, "Characteristics of surface heat flow in the South China Sea," Journal of Asian Earth Sciences, vol. 22, no. 3, pp. 265–277, 2003. ·View at Google Scholar.

23.   L. Chen and H. B. Song, "Research progress on seismic detection methods of natural gas seepage on the seabed," Natural Gas Industry, vol. 26, no. 7, pp. 35–39, 2006 (Chinese).

24.   J. X. He, B. Xia, S. L. Zhang, P. Yan, and H. L. Liu, "Origin and distribution of mud diapirs in the Yinggehai basin and their relation to the migration and accumulation of natural gas," Geology in China, vol. 33, no. 6, pp. 1336–1344, 2006 (Chinese).

25.   A. Gay, M. Lopez, P. Cochonat, M. Séranne, D. Levaché, and G. Sermondadaz, "Isolated seafloor pockmarks linked to BSRs, fluid chimneys, polygonal faults and stacked Oligocene-Miocene turbiditic palaeochannels in the Lower Congo Basin," Marine Geology, vol. 226, no. 1-2, pp. 25–40, 2006.

26.   S. Hustoft, J. Mienert, S. Bünz, and H. Nouzé, "High-resolution 3D-seismic data indicate focussed fluid migration pathways

above polygonal fault systems of the mid-Norwegian margin," Marine Geology, vol. 245, no. 1–4, pp. 89–106, 2007.

27. C. Berndt, S. Bünz, and J. Mienert, "Polygonal fault systems on the mid-Norwegian margin: a long term source for fluid flow," in Subsurface Sediment Mobilization, P. Rensbergen, R. R. Hillis, A. J. Maltman, and C. K. Morley, Eds., pp. 283–296, Geology Society of London, London, UK, 2003.

28. J. Cullen, D. C. Mosher, and K. Louden, "The mohican channel gas hydrate zone, scotian slope: geophysical structure," in Proceedings of the 6th International Conference on Gas Hydrates, Vancouver, Canada, 2008.

29. J. Watterson, J. Walsh, A. Nicol, P. A. R. Nell, and P. G. Bretan, "Geometry and origin of a polygonal fault system," Journal of the Geological Society, vol. 157, no. 1, pp. 151–162, 2000.

30. S. J. Roberts and J. A. Nunn, "Episodic fluid expulsion from geopressured sediments," Marine and Petroleum Geology, vol. 12, no. 2, pp. 195–204, 1995.

31. J. X. He, B. Xia, D. S. Sun, S. L. Zhang, and B. M. Liu, "Hydrocarbon accumulation, migration and play targets in the Qiongdongnan Basin, South China Sea," Petroleum Exploration and Development, vol. 33, no. 1, pp. 53–58, 2006 (Chinese).

32. Y. H. Zhu, B. H. Wu, X. R. Luo, and G. X. Zhang, "Geochemical characteristics of hydrocarbon gases and their origin from the sediments of the South China Sea," Geoscience, vol. 22, no. 3, pp. 407–414, 2008 (Chinese).

Chapter 2

The Dependence of Electrical Resistivity-Saturation Relationships on Multiphase Flow Instability

Zoulin Liu and Stephen M. J. Moysey

Department of Environmental Engineering and Earth Science, Clemson University, Clemson, SC 29670, USA

ABSTRACT

We investigate the relationship between apparent electrical resistivity and water saturation during unstable multiphase flow. We conducted experiments in a thin, two-dimensional tank packed with glass beads, where Nigrosine dyed water was injected uniformly along one edge to displace mineral oil. The resulting patterns of fluid saturation in the tank were captured on video using the light transmission method, while the apparent resistivity of the tank was continuously measured. Different experiments were performed by varying the water application rate and orientation of the tank to control the generalized Bond number, which describes the balance between viscous, capillary, and gravity forces that affect flow instability. We observed the resistivity index to gradually decrease as water saturation increases in the tank, but sharp drops occurred as individual fingers bridged the tank. The magnitude of this effect decreased as the displacement became increasingly unstable until a smooth transition occurred for highly unstable flows. By analyzing the dynamic data using Archie's law, we found that the apparent saturation exponent increases linearly between approximately 1 and 2 as a function of generalized Bond number, after which it remained constant for unstable flows with a generalized Bond number less than −0.106.

INTRODUCTION

Multiphase fluid flow in porous media is an important problem for applications including petroleum production [1–3], migration of nonaqueous phase liquids (NAPLs) in soils and aquifers [4–6], and CO_2 sequestration [7]. Viscous, capillary and gravity forces interact in immiscible two phase flow systems to produce stable or unstable flow regimes [8–11]. In a stable flow regime the displacement of one fluid for another will occur along a stable front. In unstable flow regimes, fingering can occur along the displacement front. As a result, the invading fluid phase can bypass significant amounts of the original fluid phase, leaving it in place in the medium.

Electrical resistivity measurements are commonly used to investigate fluid saturations in multiphase flow systems [12– 16]. The resistivity index provides an expression of resistivity for multiphase flow systems that is directly related to the degree of water saturation of the medium, S_w, through Archie's law [17]. When mineral surface conductivity is insignificant, the resistivity index I_R is equal to the ratio of the resistivity of the sample (ρ_w) measured at saturation S_w to the resistivity of the sample measured at 100% water saturation (ρ_s) (Equation (1)). The saturation exponent, n, is an empirical constant that is conceptually related to the connectivity of the electrically conductive phase, that is, water:

$$I_R = \frac{\rho_w}{\rho_s} = S_w^{-n}.$$

(1)

The saturation exponent is usually determined experimentally from measurements of I_R and S_w using (1). For example, Sweeney and Jennings [18] experimentally found the saturation exponent to be 1.61 for water wet carbonates, though this increased up to 12.27 for samples treated to become oil wet. Zhou et al. [16] used percolation models to show that the saturation exponent in strongly water wet materials is on the order of 1.9, whereas it can increase to over 3.5 for intermediate and oil wet systems. The saturation exponent is dependent on the presence of nonconductive fluid in the pore space and wettability of the rock [12, 19]. Although Archie's Law is widely used to determine fluid saturation from resistivity measurements, it is not always valid as the saturation-resistivity relationship depends on the wettability, saturation history, content of clay minerals, salinity of the brine phase and distribution of water and oil in the rock [3, 12, 15, 16]. Most experimental studies of the dependence between saturation and apparent resistivity were based on the assumption that fluid distribution within the sample was homogeneous, which can be rarely obtained during a transient displacement experiment. We experimentally investigate the influence of flow instability on apparent resistivity and the saturation exponent in Archie's law. To this end, multiphase flow experiments are conducted where water is used to displace mineral oil in a two-dimensional (2D) flow system. In this paper, we conduct a series of experiments where the stability of the flow is controlled by varying the water inflow rate and angle of the tank. Measurements of the bulk resistivity of the tank are obtained during the flow experiments. Transient estimates of average water saturation in the tank, S_w are derived from video collected using the light transmission method. Using these measurements we observe the relationship between saturation and resistivity for a range of flow conditions. We further evaluate whether there is a dependence of the saturation exponent in Archie's law on the flow conditions in a porous medium.

BACKGROUND ON FLOW INSTABIL-ITY

It is well known that variations in the magnitude and connectivity of permeability could lead to flow channeling in reservoirs and consequently to a reduction in oil production [20]. Even in a homogeneous medium, flow instability can cause viscous fingering that also increases the residual oil volume left behind in a reservoir [8, 9]. Flow instability is affected by the cumulative effects of capillary, buoyancy, and viscous forces. For example, viscous forces can destabilize the displacement front into narrow fingers if a less viscous fluid displaces a more viscous fluid, whereas gravity plays a stabilizing effect when a lighter fluid is on top of a denser phase [10]. The balance between forces in a two phase flow system can be quantified using the dimensionless Bond and capillary numbers along with the viscosity ratio. The viscosity ratio (M) is defined as the viscosity of displacing fluid μ_w divided by the viscosity of the displaced fluid μ_n. Viscous fingering can be observed when the viscosity ratio is less than 1 and the viscous force overcomes capillary and gravity effects. The Bond number (B_o), given in (2), expresses the relative importance of gravitational to capillary forces in a multiphase flow system [9, 10]. In contrast, the capillary number (C_a) in (3) expresses the balance between viscous to capillary forces [9, 10, 21]

$$B_o = \frac{\Delta P_{grav}}{\Delta P_{cap}} = \frac{\Delta \rho g a^2}{\gamma} \sin \varphi,$$

$$(2)$$

$$C_a = \frac{\Delta P_{visc}}{\Delta P_{cap}} = \frac{\mu_w v a^2}{\gamma k}.$$

$$(3)$$

In these expressions, μ_w is the viscosity of wetting fluid, v is the filtration or Darcy velocity, a is the typical pore radius, γ is surface tension, Δ_ρ is the density difference between the two fluids, g is the acceleration due to the gravity, $\square$ is the angle of flow relative to horizontal, and k is the permeability of the porous medium [9, 10]. The capillary and Bond numbers can be combined to produce the generalized Bond number (B_0^*) given in the following [9, 10]:

$$B_o - C_a = \frac{a^\sim}{vk}(\Delta\rho gk \sin\varphi -$$

(4)

The value of the generalized Bond number plays a critical role for determining the occurrence of viscous instabilities. For $B_0^* > 0$ the flow is stable and a compact and flat displacement front occurs as illustrated in Figure 1(a). However, when $B_0^* < 0$ the flow is unstable and fingering produces an uneven and often rapid movement of the infiltrating phase through the displaced phase in a porous medium (Figure 1(b); the white region in the images corresponds to oil, whereas the black areas correspond to Nigrosine dyed water). Despite the obvious contrast in macroscopic behavior, Meheust et al. note that a radical change in the local dynamics of the interface does not occur during the transition between stable and unstable displacement [10].

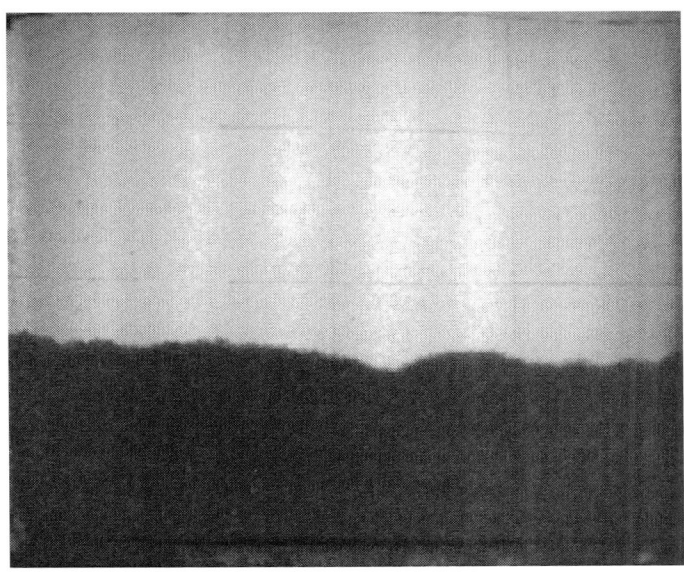

(a)

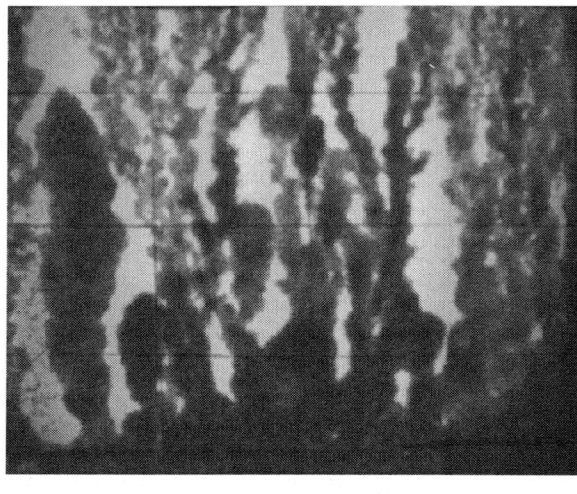

(b)

Figure 1: Comparison of (a) stable flow conditions ($B_0^* = 0.00612$) and (b) unstable flow conditions ($B_0^* = -0.248$; experiment #12 in Table 3) for oil (white) displaced by Nigrosine dyed water (black).

METHODS AND EXPERIMENTAL SETUP

The main goal of this work is to determine the relationship between the saturation exponent in Archie's law and the degree of flow instability in a porous medium as quantified by the Bond and Capillary numbers. To achieve this objective, resistivity index curves were measured during the displacement of mineral oil by water in a 2D tank packed with glass beads. Four-wire resistance measurements were collected throughout the experiment while the light transmission method was used to simultaneously monitor changes in saturation. The effect of gravity on flow instability is controlled by changing the orientation of the tank to achieve different Bond numbers. Experiments at different flow rates were conducted to control the relative importance of viscous forces by varying the capillary number.

Experimental Setup

The fluids used in these experiments are water and mineral oil (EMD Chemicals, NJ, USA). The properties of each fluid are given in Table 1. We focus on a situation where a denser fluid with low viscosity (water) displaces a less dense, more viscous fluid (mineral oil) from below leading to a low viscosity ratio (0.015). Viscous fingering is therefore possible in this system. Negrosine (Acros Organics) dye was added to the water phase to provide contrast with the clear mineral oil to allow visual tracking of the displacement front and the development of fingers. This particular dye was selected because it did not partition from the water to oil phase in initial static tests conducted in beakers. The electrical conductivity of the water used in each experiment varied from 71.8–91.5 μs/cm.

Table 1: Properties of the fluids used in the experiments

Wetting phase, water (with 0.05 g/L nigrosine)	
Density, ρ_w	1000 kg/m³
Dynamic viscosity, μ_w	N.s/m²
Non-wetting phase, mineral oil	
Density, ρ_n	880 kg/m³
Dynamic viscosity, μ_n	0.068 N.s/m²
Interfacial tension [22], γ	0.049 N/m
Viscosity Ratio, M	0.015

The experiments were conducted in the specially designed 2D acrylic tank shown in Figure 2. The dimensions of the interior flow cell of the tank are 45 cm × 40 cm × 1.25 cm. For all of the experiments in the study, the flow cell was packed with 2 mm diameter glass beads (Walter Stern). The entire cell was designed to be pressure sealed, thereby allowing for the tank to be oriented at arbitrary geometries. The outlet pressure of the tank was held at a constant positive pressure by keeping the discharge reservoir above the tank (Figure 2). Details regarding the physical properties of the tank are summarized in Table 2.

Table 2: Physical properties of the flow system

Length	40 cm
width	45 cm
Thickness	1.25 cm
Porosity, ε	0.30
Formation factor, F_f	3.04
Permeability, k	57 Darcy
Grain Size, D	2 mm

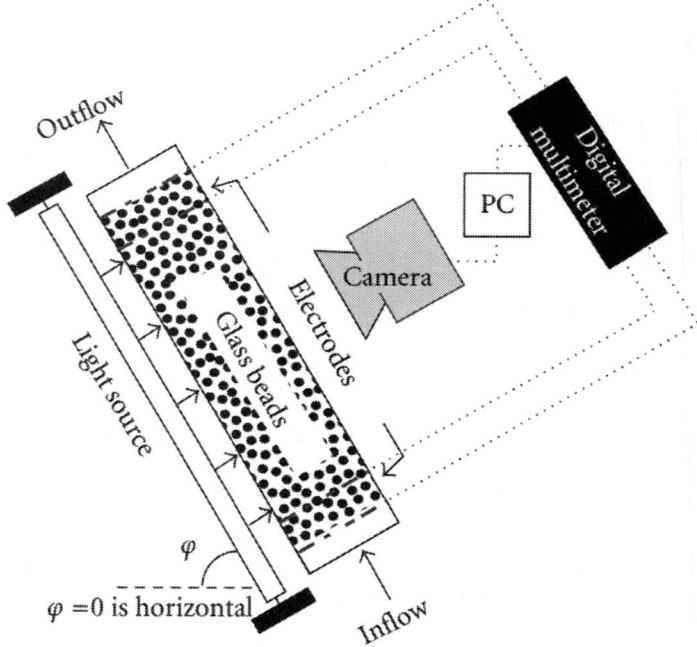

Figure 2: Side view sketch of the experimental setup for the resistivity cell with light transmission imaging and resistivity measurement systems.

The tank could tilt to arbitrary angles so as to vary the effect of gravity on flow and control the Bond number. The component of gravity acting on the flow system is determined by $g_\square = g \times \sin(\square)$, where g is acceleration due to gravity and $\square$ is the angle of the tank relative to horizontal. For each experiment water was injected into the tank at a constant rate selected to achieve a specified capillary number using a variable rate peristaltic pump (pump head: HV-07015-20, Master Flex). The displacing water phase is injected through a porous plastic plate covering the entire inlet surface of the beads to ensure the injection is uniform. The displaced oil phase is expelled from a similar outlet port at the opposite end of the flow cell. Both gravity effects, that is, Bond number, and flow rate, that is, capillary number, influence the stability of flow in our experiments and can be changed independently of each other. A complete listing of tank orientations and flow rates used in the experiments is given in Table 3..

Table 3: Summary of the tank inclination angle ($\square$), pumping rate (Q), and fluid conductivity (σ_w) for each of the 34 experiments. The corresponding characteristic numbers C_a, B_o and B_0^* are also given (pore size estimated as $0.414 \times$ grain size following [23])

Experiment index	1	2	3	4	5	6	7
Q (mL/min)	189	251	67	119	157	27	67
φ (degree)	90	90	90	90	90	90	90
σ_w (μs/cm)	98	93	90.2	93.7	81.4	126	94.7
B_o	1.65E−02	1.65E−02	1.65E−02	1.65E−02	1.65E−02	1.65E−02	1.65E−02
C_a	1.39E−01	1.85E−01	4.94E−02	8.78E−02	1.16E−01	1.99E−02	4.94E−02
B_0^*	−1.23E−01	−1.69E−01	−3.30E−02	−7.13E−02	−9.94E−02	−3.47E−02	−3.30E−02

Experiment index	8	9	10	11	12	13	14
Q (mL/min)	119	27	189	251	358	67	99
φ (degree)	90	90	90	90	90	90	90
σ_w (μs/cm)	82	78.7	79.5	76.7	78.4	80.1	83.1
B_o	1.65E−02	1.65E−02	1.65E−02	1.85E−02	1.65E−02	1.65E−02	1.65E−02
C_a	8.78E−02	1.99E−02	1.39E−01	1.85E−01	2.64E−01	4.94E−02	7.30E−02
B_0^*	−7.13E−02	−3.47E−03	−1.23E−01	−1.69E−01	−2.48E−01	−3.30E−02	−5.66E−02

Experiment index	15	16	17	18	19	20	21
Q (mL/min)	89	52	146	119	67	27	67
φ (degree)	90	90	90	90	90	90	90
σ_w (μs/cm)	74.8	77.6	74.2	79	77.6	78.3	87
B_o	1.65E−02	1.65E−02	1.65E−02	1.65E−02	1.65E−02	1.65E−02	1.65E−02
C_a	6.57E−02	3.84E−02	1.08E−01	8.78E−02	4.94E−02	1.99E−02	4.94E−02
B_0^*	−4.92E−02	−2.19E−02	−9.13E−02	−7.13E−02	−3.30E−02	−3.47E−03	−3.30E−02

Experiment index	22	23	24	25	26	27	28
Q (mL/min)	189	251	67	67	67	67	67
φ (degree)	90	90	30	30	30	0	45
σ_w (μs/cm)	71.8	72.8	78	91.5	90	81.1	80.5
B_o	1.65E−02	1.65E−02	8.23E−03	8.23E−03	8.23E−03	0.00E+00	1.16E−02
C_a	1.39E−01	1.85E−01	4.94E−02	4.94E−02	4.94E−02	4.94E−02	4.94E−02
B_0^*	−1.23E−01	−1.69E−01	−4.12E−02	−4.12E−02	−4.12E−02	−4.94E−02	−3.78E−02

Experiment index	29	30	31	32	33	34
Q (mL/min)	67	67	67	67	358	358
φ (degree)	15	60	60	60	90	90
σ_w (μs/cm)	82.8	84	79.2	84.7	89.1	88.5
B_o	4.26E−03	1.42E−02	1.42E−02	1.42E−02	1.65E−02	1.65E−02
C_a	4.94E−02	4.94E−02	4.94E−02	4.94E−02	2.64E−01	2.64E−01
B_0^*	−4.52E−02	−3.52E−02	−3.52E−02	−3.52E−02	−2.48E−01	−2.48E−01

The bulk DC resistance of the tank was determined using the four-electrode method [4, 24]. Two pieces of copper mesh were anchored across the inflow and outflow sides of the tank to act as current electrodes. Two additional copper strips were positioned 2 cm away from each potential electrode within the tank to act as potential electrodes (Figure 2). A National Instruments PXI system with 7.5-digit digital multimeter and multiconfiguration matrix module (NI PXI-4071, PXI-2530) were used to measure the DC resistivity of the tank while switching the polarity of the current electrodes to avoid electrode polarization effects. Prior to running the flow experiments, the flow cell was filled with a saline solution and measurements were taken to calibrate the geometric factor relating tank bulk resistance to resistivity. After packing the tank with the glass beads the formation factor in Archie's law was determined to be 3.04 for our experiments by measuring resistance for several different solution conductivities. Surface conductivity effects for these large glass beads are assumed to be negligible given insignificant imaginary conductivity responses measured by spectral induced polarization [25].

The system developed for the light transmission measurements [6, 26, 27] contains a light source and detector (Figure 2). In this experiment a scientific digital camera (DFK 41BU02.H USB CCD, Imaging Source) with a 5 mm lens (H0514-MP, Imaging Source) is used as the detector to quantify the intensity of light transmitted through the tank. This camera has resolution of 1280 × 960 pixels for 32 bit images, which provides a spatial resolution of 0.39 mm per pixel or about 2 pixels per pore for a distance between the camera and the tank of 40 cm. The camera is controlled by a host computer using a LabView (National Instruments) program to obtain images at a specific frame rate during the fluid displacement. The pictures that the camera takes are in raw bmp format with no compression. Images are later converted into gray scale and analyzed using MATLAB. The light transmitted through the tank is generated by an array of fluorescent bulbs (13 W each, Bi-Pin, MA) mounted to the back of the tank in a manner allowing it to move with the tank when the experimental angle is changed.

The background reference image obtained before water is injected is subtracted from each subsequent image to overcome problems related to variations in light intensity due to the specific arrangement of light bulbs in the array. The intensity ($I_0 - I$) of the corrected image was found to have a linear relationship with the water saturation S_w

inside the porous medium as shown in the following:

$$= \frac{[1.0759 \times (I_0 - I) - 3.35}{100} \tag{5}$$

This equation is obtained from calibration experiments using a small chamber with the same material, thickness, and packing of glass beads to obtain a porosity of about 0.30, consistent with the flow cell. With the experimental setup described above, 2 main series of 34 experiments are conducted: one series is at constant Bond number of 0.0165 and the other is at constant capillary number of 0.0494. Table 3 summarizes the experimental conditions for all experiments in terms of the orientation of the tank ($\square$), pumping rates (Q), water conductivity (σ), and the corresponding Capillary number (C_a), Bond number (B_o), and generalized Bond number B_0^*.

RESULTS

The range of the Bond numbers that can be achieved by rotating the tank, that is, 0 to 0.0165, is smaller than the range of capillary numbers that can be achieved by changing the flow rate, that is, 0 to 0.264. Therefore, we can obtain the largest range of generalized Bond numbers by changing flow rate. The maximum generalized Bond number used in the experiments is −0.0035 because the digital multimeter was not able to read the high resistivity of the mineral oil in completely stable situations where water uniformly displaced the oil. The lowest (i.e., most negative) generalized Bond number investigated is −0.248 as the medium tended to compact under high internal pressures if higher flow rates were applied in the closed cell.

Saturation and Resistivity Index

The average water saturation and resistivity index of the tank over time are shown in Figure 3 for different values of the generalized Bond number. Saturations change relatively smoothly in most cases as water displaces the oil. Differences between the curves are apparent, but trends for different generalized Bond numbers are not clear. In contrast, the resistivity index curves show a distinctive change in behavior with generalized Bond number. For small values of B_0^*, that is, values near zero where flow is more stable, the resistivity index curves show large, sudden drops. In contrast, for large negative values of B_0^*, in which case the flow is highly unstable with many thin fingers formed, the reduction in resistivity index over time is smooth and regular. This result is indicative of the high sensitivity of resistivity measurements to the geometry of the water phase in the medium. Note that we use resistivity index here since the fluid resistivity varied between some of the experiments (Table 3).

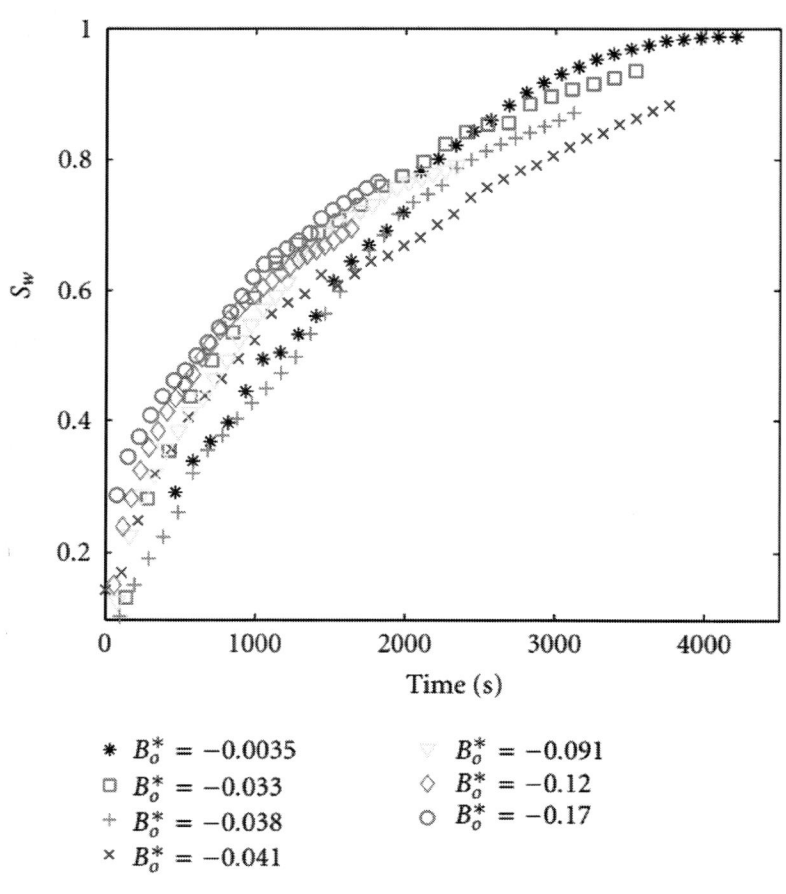

$*\ B_o^* = -0.0035$	$\triangledown\ B_o^* = -0.091$
$\square\ B_o^* = -0.033$	$\diamond\ B_o^* = -0.12$
$+\ B_o^* = -0.038$	$\circ\ B_o^* = -0.17$
$\times\ B_o^* = -0.041$	

(a)

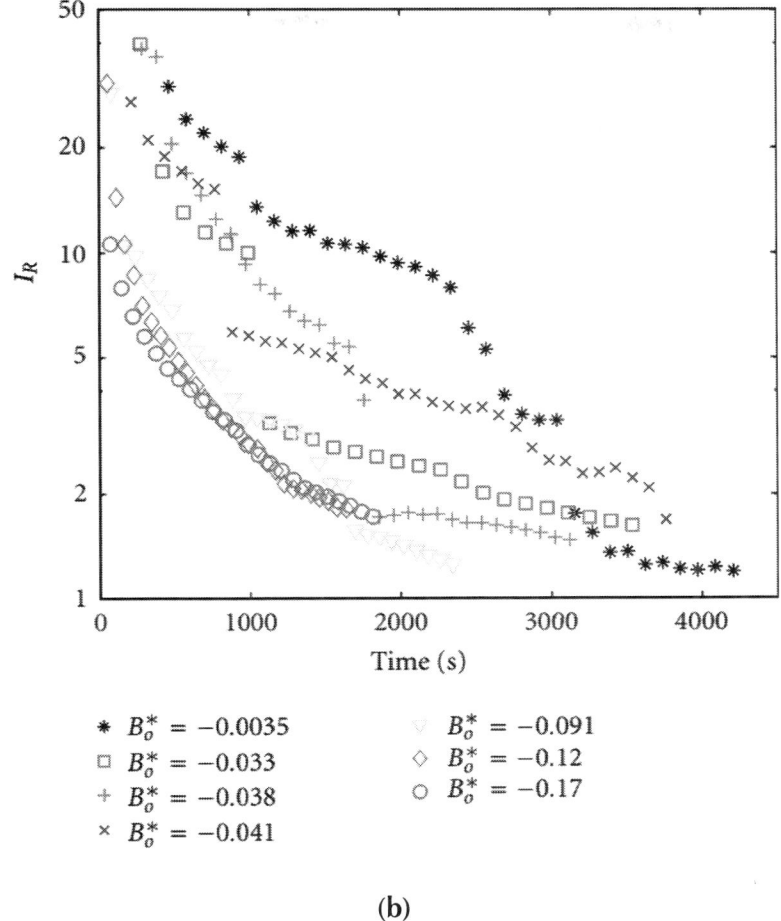

* $B_o^* = -0.0035$	$B_o^* = -0.091$
□ $B_o^* = -0.033$	◇ $B_o^* = -0.12$
+ $B_o^* = -0.038$	○ $B_o^* = -0.17$
× $B_o^* = -0.041$	

(b)

Figure 3: Changes in (a) average tank saturation and (b) resistivity index through time for varying values of B_0^* (data shown for experiment index = 9, 7, 28, 25, 17, 10, 11).

At small negative values of B_0^* the flow is stable and the water advances either as a uniform front or as large, individual fingers. Using the video collected during the experiment, each jump in resistivity index can be correlated with the time at which a finger of water reaches the tank's upper current electrode, thereby completing a new pathway for current to flow through the medium. At large negative values of B_0^*

the flow is highly unstable, producing many thin fingers. The fingers tend to reach the tank outflow in a more uniform manner, producing the relatively smooth change in resistivity observed in Figure 3(b). The patterns of fingering observed in our experiments (Figure 4), that is, increasing number of fingers and decreasing finger thickness with increasing generalized Bond number, is consistent with observations from experiments by Løvoll et al. [9] and Meheust et al. [´10] though these authors did not measure resistivity..

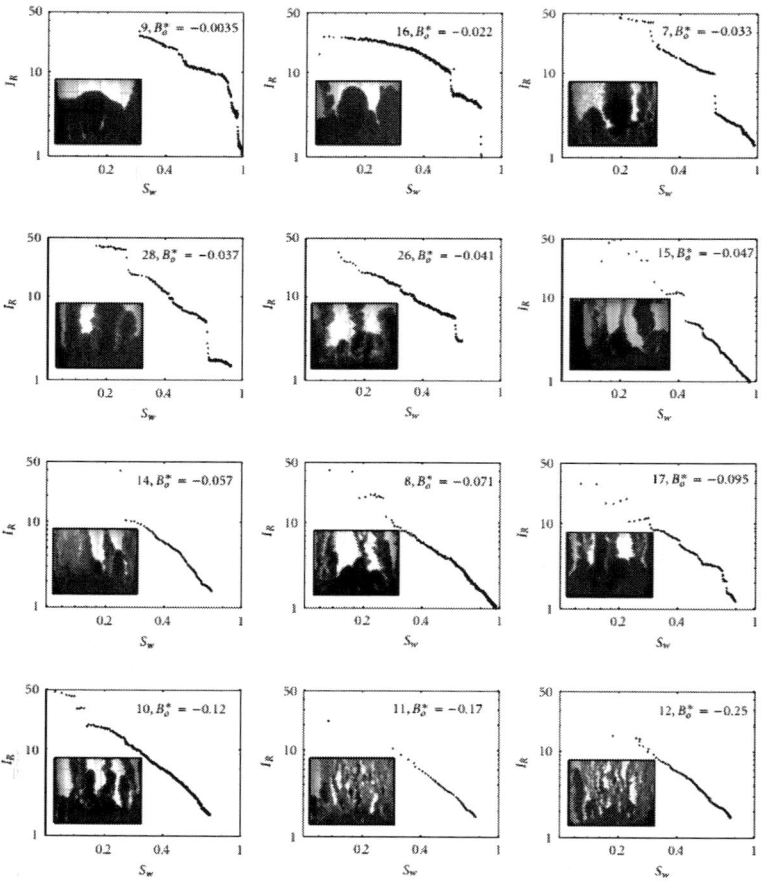

Figure 4: Dependence of resistivity index on saturation for different values of B∗ o (the experimental index and B_0^* are given in each figure).

A representative image of the flow conditions during each experiment is shown with the inset picture.

Based on experimental data, Méheust et al. [10] suggest a power law with an exponent of −0.55 to relate the front width of the fingering and the generalized Bond number:

$$W = B_o^{*-0.55},$$

(6)

where W is the measured front width of the finger, that is, the root mean square maximum extension perpendicular to the flow direction. Therefore bigger discrete drops in the resistivity index at small values of the generalized Bond number can be attributed to wider fingers reaching the electrodes and causing a larger portion of the flow cell to connect the electrodes. As the flow grows increasingly unstable, the conductive fingers have a smaller width and are more uniformly distributed in the tank, though individual fingers may form at different times. As a result, the resistivity change caused by an individual finger is small and the observed change in resistivity is smoother as a result of the progressive arrival of individual conductive fingers.

Figures 4 and 5 show the resistivity index versus water saturation for a subset of the generalized Bond numbers used in the experiments. For a given saturation value the resistivity index is generally lowest for unstable flow scenarios, that is, in Figure 5 the value of the generalized Bond number of the curves decreases (becomes more negative) from the top curve to the bottom. The slope of the resistivity index curves is flattest for small negative values of the generalized Bond number with changes in resistivity occurring as sharp drops when individual fingers reach the outflow end of the tank. As discussed previously, the magnitude of these drops in the resistivity index decrease as the generalized Bond number becomes more negative and the drops disappear when the flow is very unstable at $B_0^* = -0.123$ (Figure 5).

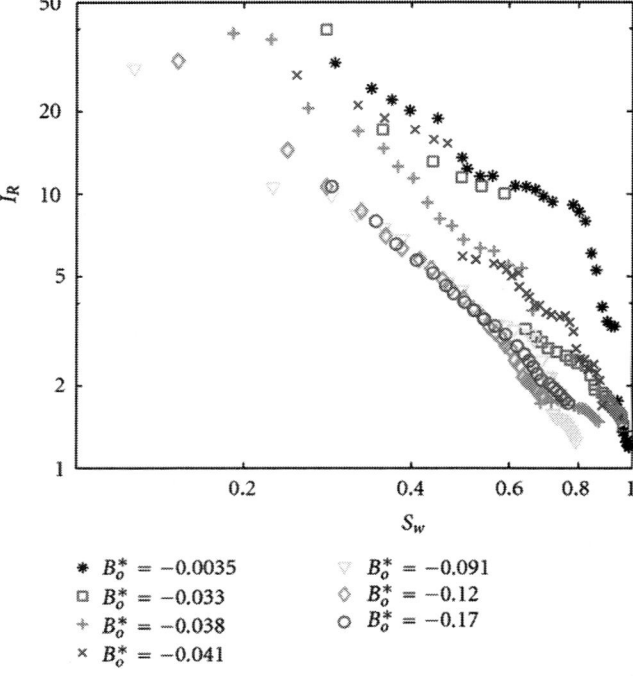

Figure 5: Direct comparison of resistivity index changes with saturation for different generalized Bond numbers.

It is notable that the resistivity index versus saturation curves for B_0^* = −0.123 (Figure 5, diamond) and $B_0^* = −0.169$ (Figure 5, circle) overlap with each other, suggesting that these two unstable flow scenarios have a similar electrical behavior. In contrast, at higher generalized Bond numbers, for example, $B_0^* = −0.00347$, the flow is more stable and a very sharp drop in I_R is observed at the end of the experiment. This rapid change in resistivity index occurs because the front of the water moving to the end the tank is relatively flat (i.e., lacks fingers) compared to the other experiments. In this case, both the water saturation and resistivity index are approximately equal to 1 at the end of the experiment, indicating that the porous medium is almost fully saturated with water (i.e., high sweep efficiency) and that the resistivity is approximately equal that observed for the 100% water saturated tank.

Influence of Flow Instability on the Saturation Exponent

Archie's law (equation (1)) is typically assumed to apply to situations where fluids are distributed uniformly throughout a porous medium. We feel it is also valuable, however, to investigate how flow instability could affect the apparent properties of the formation during unstable flow. In Archie's law the logarithms of the resistivity index (I_R) and saturation are linearly related with a slope equivalent to the negative of the saturation exponent (n). We analyze the saturation and resistivity data obtained here in a similar way to obtain an apparent saturation exponent for each experiment. For some experiments, however, the slope of these curves varies as a function of saturation. This effect is at least partially a result of the resistivity drops discussed earlier, which are a consequence of the fact that the measurements represent a dynamic flow system where preferential flows are established within the spatially finite bounds of the tank. The saturation exponent was therefore estimated for distinct sections of the resistivity-saturation curve, ignoring the sudden drops in resistivity caused by the breakthrough of individual fingers of water (Figure 6). We acknowledge that the value of the apparent saturation exponent obtained in this way may not have the same physical meaning as in typical applications of Archie's law. Regardless, this approach still provides a way to summarize the experimental data in a way that allows for effective comparison between the experiments.

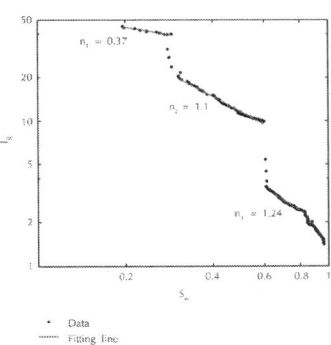

Figure 6: Illustration of how the saturation exponent was determined for experiments with discrete resistivity drops (data shown for experiment index 7).

Table 4 shows the apparent saturation exponent estimated for each experiment, averaging sections of the curves for the experiments found to have large, sudden resistivity drops. We take the average of these values to estimate the saturation exponent corresponding to a given generalized Bond number. For the unstable flow experiments, that is,

large negative values of B_0^*, the resistivity change is continuous so the saturation exponent is estimated by fitting the data with a single slope. We can therefore evaluate how the apparent saturation exponent in our experiments varies as a function of generalized Bond number using the measured resistivity index versus water saturation curves (i.e., Figures 4 and 5).

Table 4: Saturation exponent determined for each experiment

B_0^*	Data					Mean	
$B_0=0.017$, Varying C_a							
−0.0035	Experiment Index	6[1]	9[1]	20		0.70	0.11
	n	0.59	0.80	0.71			
−0.022	Experiment Index	16				0.91	—
	n	0.91					
−0.033	Experiment Index	3	13[1]	7[1]	19	0.96	0.08
	n	0.96	0.90	0.90	1.08		
−0.049	Experiment Index	15				1.30	—
	n	1.30					
−0.057	Experiment Index	14				1.51	—
	n	1.51					
−0.071	Experiment Index	4	8	18[1]		1.73	0.27
	n	1.73	1.99	1.46			
−0.091	Experiment Index	17				1.43	—
	n	1.43					
−0.099	Experiment Index	5				1.71	—
	n	1.71					

−0.12	Experiment Index	1¹	10	22		1.96	0.10
	n	2.06	1.97	1.86			
−0.17	Experiment Index	2	11	23		1.93	0.02
	n	1.92	1.96	1.92			
−0.25	Experiment Index	12	33	34		1.94	0.02
	n	1.93	1.96	1.94			
$C_a=0.049$, Varying , B_0 Transition Zone							
−0.035	Experiment Index	30	31	32		1.38	0.18
	n	1.41	1.17	1.54			
−0.038	Experiment Index	28				1.22	—
	n	1.22					
−0.041	Experiment Index	24	25	26		1.28	0.19
	n	1.10	1.47	1.27			
−0.045	Experiment Index	29				1.40	—
	n	1.40					
−0.049	Experiment Index	27				1.59	—

¹Significant resistivity drops observed in the data.

We find that the apparent saturation exponent increases when flow becomes increasingly unstable, that is, B_0^* becomes more negative (Figure 7). The saturation exponent reaches a constant value of 1.94 when generalized Bond number is less than −0.106; notably, this is consistent with the saturation index estimated by Zhou et al. [16] for strongly water wet materials using percolation theory. We quantify the relationship observed in Figure7 between generalized Bond number and the saturation exponent as follows:

$$9.3 \times B_0^* + 0.9 \quad \text{for} \quad -0.106 < B_0^* < -0$$

$$n = 1.94 \quad \text{for } B_0^* < -0.106. \tag{7}$$

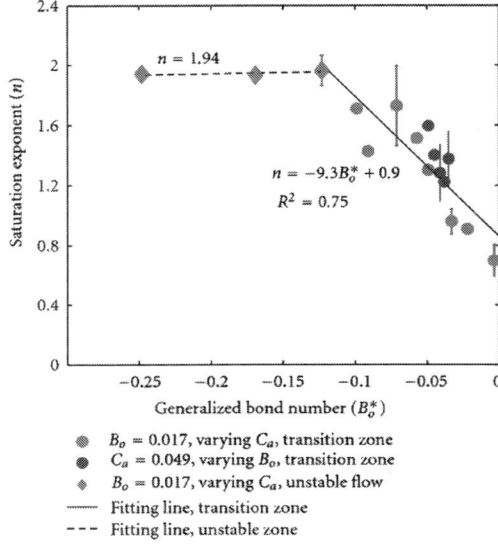

Figure 7: Saturation exponent as a function of generalized bond number B_0^* for all experiments with a linear fit over the range $-0.106 < B_0^* < -00347$. Error bars indicate the standard deviation in the saturation exponent estimate from experiments performed with the same value of B_0^*.

The observed dependence of the saturation exponent on generalized Bond number demonstrates that relationships used to estimate fluid saturation from resistivity measurements, for example, Archie's law, must be dynamic and take into account the way in which a reservoir is managed and produced. The saturation exponent is fundamentally related to the geometry of the conductive and nonconductive phases within a porous medium, specifically the connectivity of the conductive phase. Given that instability has an overwhelming influence on fluid distributions during multiphase flow, this process will also strongly influence the saturation index. We identify the increasing connectivity of the water phase between the current electrodes as the primary cause for dependence of the saturation exponent on the generalized Bond number for $-0.106 < B_0^* < -0.00347$. However, when the system reaches a certain minimum degree of connectivity due to flow instability, that is, when $B_0^* < -0.106$, there no longer appears to be a dependence between the saturation exponent and generalized Bond number. These results suggest that geophysicists must collaborate with petroleum engineers to understand the dynamics of a given flow system before attempting to estimate saturation from resistivity measurements.

CONCLUSIONS

The influence of flow instability on electrical resistivity measurements was investigated by displacing a light, but viscous mineral oil by water in a homogeneous porous medium. The experimental setup allowed the effects of gravity and flow rate to be controlled, thereby permitting the generalized Bond number to be changed between different experiments. Video showing the distribution of the fluid phases in the tank allowed us to determine the overall tank saturation throughout the experiment, while continuous measurements of average tank resistivity were also collected throughout each experiment.

In the resulting data we observed a transition between stable and unstable displacement. By analyzing the saturation and resistivity data using Archie's law, we found that the resulting apparent saturation exponent depends linearly on the generalized Bond number until it reaches a maximum of 1.94 for highly unstable flow systems, that is,

when $B_0^* < -0.106$. These results suggest that the interpretation of fluid saturation from electrical resistivity measurements must take into account flow conditions within the subsurface, since the onset of flow instability is a primary control on the spatial distribution of the fluids in the subsurface.

ACKNOWLEDGMENTS

Acknowledgment is made to the Donors of the American Chemical Society Petroleum Research Fund for support (or partial support) of this research.

REFERENCES

1. K. Mogensen, E. H. Stenby, and D. Zhou, "Studies of waterflooding in low-permeable chalk by use of X-ray CT scanning," Journal of Petroleum Science and Engineering, vol. 32, no. 1, pp. 1–10, 2001.

2. D. Tiab and E. C. Donaldson, Petrophysics—Theory and Practice of Measuring Reservoir Rock and Fluid Transport Properties, Gulf Professional, 2nd edition, 2004.

3. E. Toumelin and C. Torres-Verdín, "Object-oriented approach for the pore-scale simulation of DC electrical conductivity of two-phase saturated porous media," Geophysics, vol. 73, no. 2, pp. E67–E79, 2008.

4. C. Aggelopoulos, P. Klepetsanis, M. A. Theodoropoulou, K. Pomoni, and C. D. Tsakiroglou, "Large-scale effects on resistivity index of porous media," Journal of Contaminant Hydrology, vol. 77, no. 4, pp. 299–323, 2005.

5. T. W. J. Bauters, D. A. DiCarlo, T. S. Steenhuis, and J. Y. Parlange, "Preferential flow in water-repellent sands," Soil Science Society of America Journal, vol. 62, no. 5, pp. 1185–1190, 1998.

6. N. Weisbrod, M. R. Niemet, and J. S. Selker, "Imbibition of saline solutions into dry and prewetted porous media," Advances in Water Resources, vol. 25, no. 7, pp. 841–855, 2002.

7. N. B. Christensen, D. Sherlock, and K. Dodds, "Monitoring CO_2 injection with cross-hole electrical resistivity tomography," Exploration Geophysics, vol. 37, pp. 44–49, 2006.

8. G. M. Homsy, "Viscous fingering in porous-media," Annual Review of Fluid Mechanics, vol. 19, pp. 271–311, 1987.

9. G. Løvoll, Y. Méheust, K. J. Måløy, E. Aker, and J. Schmittbuhl, "Competition of gravity, capillary and viscous forces during drainage in a two-dimensional porous medium, a pore scale study," Energy, vol. 30, no. 6, pp. 861–872, 2005.

10. Y. Méheust, G. Loøvoll, K. J. Måløy, and J. Schmittbuhl, "Interface scaling in a two-dimensional porous medium under combined viscous, gravity, and capillary effects," Physical Review E, vol. 66, no. 5, Article ID 051603, 12 pages, 2002.

11. J. P. Stokes, D. A. Weitz, J. P. Gollub et al., "Interfacial stability of immiscible displacement in a porous medium," Physical Review Letters, vol. 57, no. 14, pp. 1718–1721, 1986.

12. W. G. Anderson, "Wettability literature survey-part 3: the effect of wettability on the electrical properties of porous media," Journal of Petroleum Technology, vol. 39, no. 13, pp. 1371–1378, 1986.

13. S. Bekri, J. Howard, J. Muller, and P. M. Adler, "Electrical resistivity index in multiphase flow through porous media," Transport in Porous Media, vol. 51, no. 1, pp. 41–65, 2003.

14. M. J. Blunt, M. D. Jackson, M. Piri, and P. H. Valvatne, "Detailed physics, predictive capabilities and macroscopic consequences for pore-network models of multiphase flow," Advances in Water Resources, vol. 25, no. 8–12, pp. 1069–1089, 2002.

15. A. K. Moss, X. D. Jing, and J. S. Archer, "Wettability of reservoir rock and fluid systems from complex resistivity measurements," Journal of Petroleum Science and Engineering, vol. 33, no. 1–3, pp. 75–85, 2002.

16. D. Zhou, S. Arbabi, and E. H. Stenby, "A percolation study of wettability effect on the electrical properties of reservoir rocks," Transport in Porous Media, vol. 29, no. 1, pp. 85–98, 1997.

17. G. E. Archie, "The electrical resistivity log as an aid in determining some reservoir characteristics,"Petroleum Transactions of AIME, vol. 146, pp. 54–62, 1942.

18. S. A. Sweeney and H. Y. Jennings, "Effect of wettability on the electrical resistivity of carbonate rock from a petroleum reservoir," Journal of Physical Chemistry, vol. 64, no. 5, pp. 551–553, 1960.

19. D. Abdassah, P. Permadi, Y. Sumantri, and R. Sumantri, "Saturation exponent at various wetting condition: fractal modeling of thin-sections," Journal of Petroleum Science and Engineering, vol. 20, no. 3-4, pp. 147–154, 1998.

20. J. M. Hovadik and D. K. Larue, "Static characterizations of reservoirs: refining the concepts of connectivity and continuity," Petroleum Geoscience, vol. 13, no. 3, pp. 195–211, 2007.

21. R. Lenormand, "Liquids in porous media," Journal of Physics, vol. 2, pp. SA79–SA88, 1990.

22. H. Yoon, M. Oostrom, and C. J. Werth, "Estimation of interfacial tension between organic liquid mixtures and water," Environmental Science and Technology, vol. 43, no. 20, pp. 7754–7761, 2009.

23. J. F. Villaume, "Investigations at sites contaminated with dense, non-aqueous phase liquids (NAPLs),"Ground Water Monitoring Review, vol. 5, no. 2, pp. 60–74, 1985.

24. G. F. Tagg, "Practical investigations of the earth resistivity method of geophysical surveying,"Proceedings of the Physical Society, vol. 43, no. 3, pp. 305–323, 1931.

25. N. Hao, J. Waterman, T. A. Kendall, S. M. Moysey, and D. Ntarlagiannis, "Resolving IP mechanisms using micron-scale surface conductivity measurements and column SIP data," Geochimica et Cosmochimica Acta, vol. 74, pp. A380–A380, 2010.

26. C. J. G. Darnault, J. A. Throop, D. A. Dicarlo, A. Rimmer, T. S. Steenhuis, and J. Y. Parlange, "Visualization by light transmission of oil and water contents in transient two-phase flow fields," Journal of Contaminant Hydrology, vol. 31, no. 3-4, pp. 337–348, 1998.

27. M. R. Niemet and J. S. Selker, "A new method for quantification of liquid saturation in 2D translucent porous media systems using light transmission," Advances in Water Resources, vol. 24, no. 6, pp. 651–666, 2001.

Chapter 3

Geochemical Analysis as a Complementary Tool to Estimate the Uplift of Sediments Caused by Shallow Gas Hydrates in Mounds at the Seafloor of Joetsu Basin, Eastern Margin of the Japan Sea

Antonio Fernando Menezes Freire[1], Ryo Matsumoto[2], and Fumio Akiba[3]

[1]Petrobras Research Center (CENPES), Department of Geochemistry, Avenida Horacio Macedo 950, Cidade Universit ´ aria, Ilha do. ´ Fundao, 21941-915 Rio de Janeiro, RJ, Brazil

[2]Meiji University, 1-1-1 Higashi-Mita, Tama-ku Kawasaki, Kanagawa 214-8571, Japan

[3]Diatom Minilab Akiba Ltd., 632-12 Iwasawa, Hanno, Saitama 357-0023, Japan

ABSTRACT

The Holocene sediments of the eastern margin of the Japan Sea are characterized by high total organic carbon (TOC) and total nitrogen (TN) contents, low TOC/TN and TS/TOC values with enriched $\delta^{13}C_{org}$ signatures, as a result of high marine productivity during present oxic highstand. On the other hand, the LGM sediments are characterized by low TOC and TN contents, high TOC/TN and TS/TOC values with depleted $\delta^{13}C_{org}$ signatures, characteristic of C3-derived terrestrial organic matter input during that anoxic lowstand. However, at the top of mounds at the seafloor, where gas hydrate and authigenic carbonate nodules occur, the host sediments have a mixture of both Holocene and LGM geochemical signatures. Both gas hydrate and authigenic carbonate, formed by the anaerobic oxidation of methane, increased the sedimentary volume and caused an uplift of older sediments, inducing mound formation. The thickness of the Holocene sediments over mounds is very small or absent exposing the last glacial maximum (LGM) sediments to the seafloor. The uplift of the LGM sediments within mounds is estimated to be >2 m. We conducted geochemical analysis to detect such sediment movement, using samples collected by shallow cores in the Joetsu Basin, eastern margin of the Japan Sea.

INTRODUCTION

Japan Sea is a typical back-arc basin formed behind the Japanese island-arc system initiated by the rifting of the eastern margin of the Eurasian Continent at around 25 Ma [1]. The opening was almost completed before 15 Ma [2, 3]. During the Middle Pliocene, the tectonic style changed from the extensional to compressive [4] and a series of NNE-SSW trending folds were formed along the eastern margin of the Japan Sea [5], where an incipient subduction zone extends throughout the western side of the Japanese island-arc system [6]. As a result, several potential hydrocarbon traps were formed during this period and continuous subsidence created kitchen areas with mature source rocks [5]. Joetsu Basin is located southwest of Sado Island (Figure 1) and was formed during the Miocene [5, 7]. A favorable source rock was developed by high production of organic matter under anoxic conditions in the Nanatani Fm. (<12.5 Ma) and Teradomari Fm. (12.5~5.5 Ma) [5]. Oil generation occurred in the Miocene and a 15 m oil column was confirmed in tuffaceous sandstones in the lower part of the Shiiya Fm. (5.5~3.5 Ma) [5]. The top of the Nishiyama Fm. (3.5~1.3 Ma) is characterized by a "domino style" with several horsts and grabens and both normal and reverse faults are observed, reflecting the complex stress field involved [8]. Some of these faults belong to the rifting time and were reactivated during the inversion tectonic process [9]. The Haizume formation has been deposited since the late Pleistocene, and it is dominated mainly by clayey sediments [10].

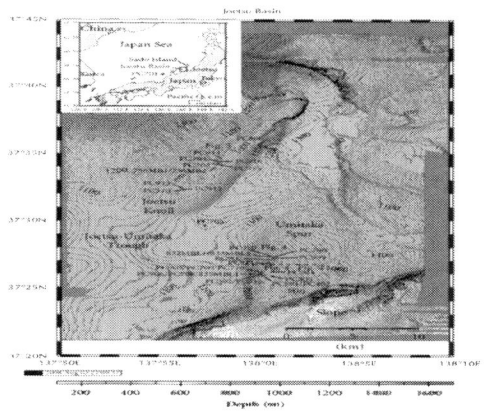

Figure 1: Location map of Joetsu Basin, Umitaka Spur, and Joetsu Knoll. Core locations are also plotted. Red lines indicate the position of the next figures.

Umitaka Spur and Joetsu Knoll are two anticlines formed since the middle Pliocene with a regional NE-SW trend [11], located approximately 30 Km offshore Joetsu city (Figure 1). A complex NE-SW axial fault system, composed of both normal and reverse faults, is observed in the central parts of both anticlines [11, 13]. The combination of faults, carrier beds, depositional surfaces, thermogenic gas source location, and the geometry of the anticlines focuses gas migration towards the top of both structures and also provides gas to the gas hydrate stability zone (GHSZ). The axial fault system results in an inefficient seal and trap and high amounts of methane are delivered to the GHSZ, where gas hydrate is generated within gas chimneys (Figure 2) [11, 13].

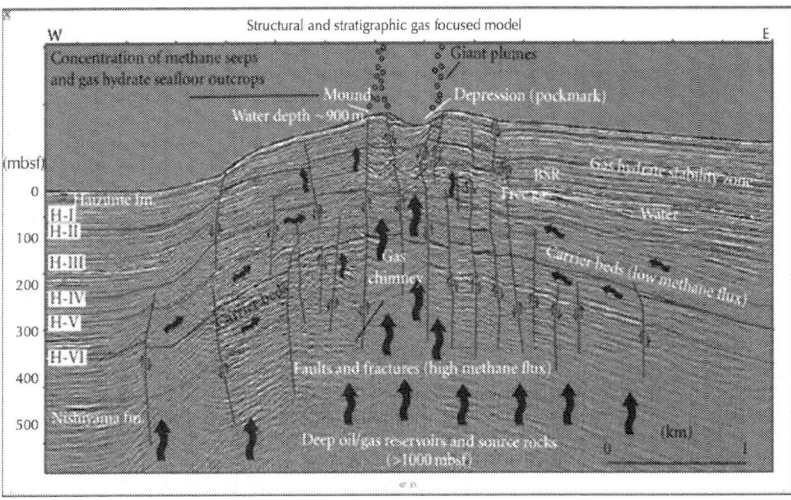

Figure 2: Structural-stratigraphic control on the gas hydrates of Umitaka Spur. Arrow size suggests the intensity of gas migration. The combination of the shape of the anticline, carrier beds, and the axial fault system induces gas migration to the top of the structure supplying gas for the GHSZ and to the water column forming giant plumes. See profile location in Figure 1. This model was modified from Freire et al. [11].

Associated to the gas chimneys, Matsumoto [12, 15] observed several mounds and pockmarks at the seafloor. According to his studies, mounds were formed by the crystallization of gas hydrates, while pockmarks were the result of intense gas hydrate dissociation during the LGM, when the sea level dropped ca. 120 m below that of the present [16]. He proposed that pockmarks were the final stage after mound formation, when gas hydrate stability conditions were severely lost during the lowstand of the LGM.

The main propose of this study is to detail the influence of both gas hydrate and authigenic mineral formation for the growth of mounds related to gas seepages. Both processes within the porous space increase the sedimentary volume and induce the formation of mounds by an uplift of old sediments associated with high methane flux venting. To confirm this, we performed core descriptions and geochemical analysis in sediments collected by piston and push cores at the mounds and surrounding areas for comparing their geochemical signatures.

METHODOLOGY

Dozens of piston and push cores were collected for gas hydrate research from the Joetsu Knoll, Umitaka Spur, and surrounding areas since 2005 by the R/V Umitaka Maru of the Tokyo University of Marine Science and Technology and by the R/V Kaiyo of the Japan Agency for Marine-Earth Science and Technology (JAMSTEC). These studies have been conducted by the University of Tokyo and other institutions, providing a significant improvement in the geological knowledge of the eastern margin of the Japan Sea, in particular of the Joetsu Basin [11–15, 17–19].

A total of 22 piston and push cores, respectively, 6 to 9 m and 0.50 m long, were used for this study, recovered on both Umitaka Spur and Joetsu Knoll at Joetsu Basin (Figure 1). One piston core (PC701) was collected at Oki Trough (Figure 1), representing the reference characteristics for open sea conditions, where there is no indication of the occurrence of gas hydrates. From these cores, a total of 475 samples were analyzed for TC, TOC, TN, TS, and $^{13}C_{org}$. A total of 67 samples belonging to the gas hydrate-bearing sediments (Figure 3) were collected over topographic mounds (Figures 4, 5, and 6).

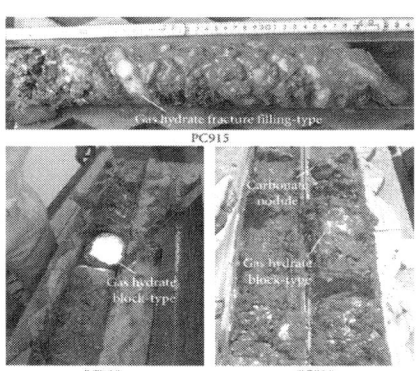

Figure 3: Gas hydrate-bearing sediments collected by piston cores in Joetsu Basin. Blocky- and fracture-filling gas hydrate can be observed. Both non-disturbed and soupy sediments disturbed during piston coring were sampled and analyzed to compare with nondisturbed sediments collected far from mounds.

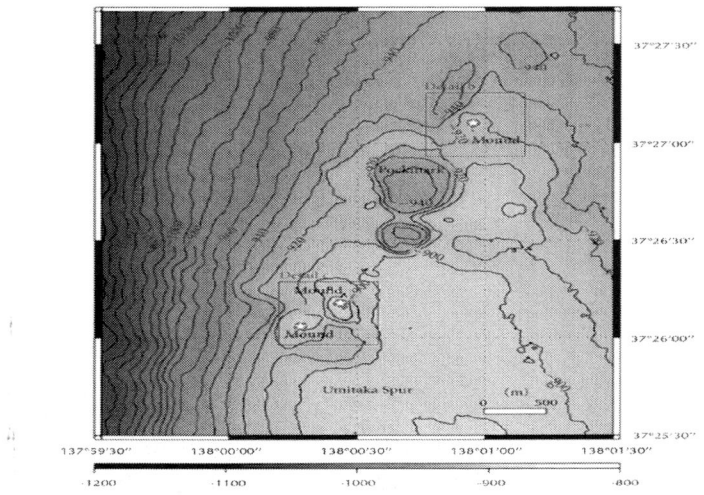

Figure 4: Map of Umitaka Spur. Red stars represent giant plumes and methane hydrate exposed at the seafloor. Red squares indicate detailed images of next figure. Mounds and pockmarks are indicated. See Figure 1 for Figure 4 location0.

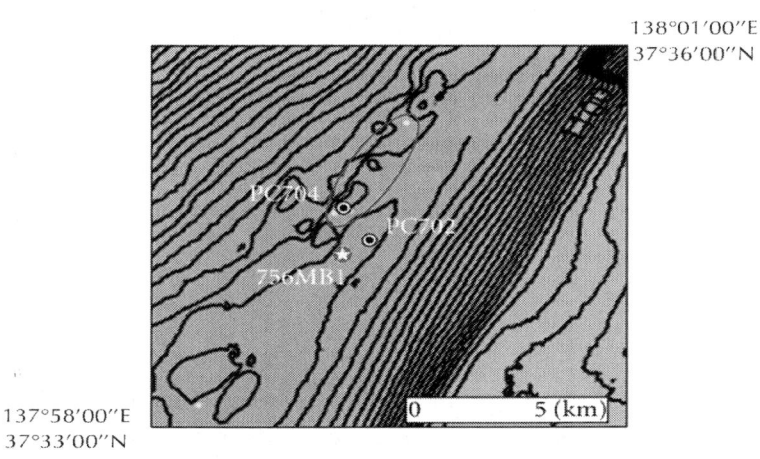

(a)

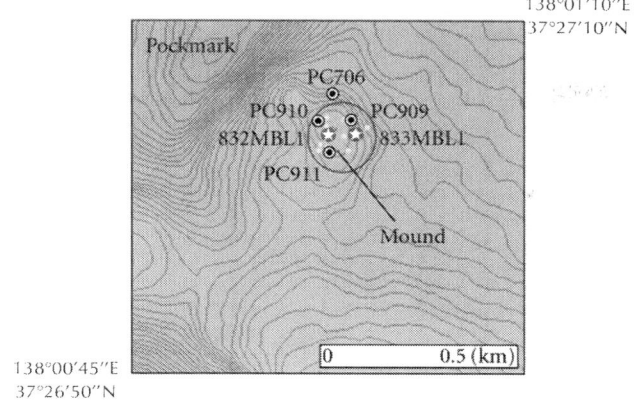

(b)

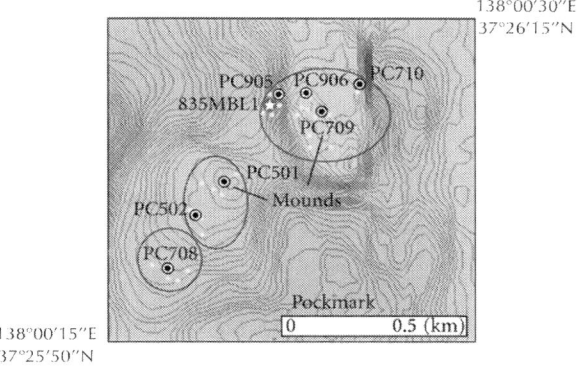

(c)

Figure 5: Detailed maps of seep sites. (a) Joetsu Knoll center, (b) Umitaka Spur north, and (c) Umitaka Spur center. See Figures 1 and 4 for map locations. Yellow circles indicate giant plumes observed by echosounder during the last few years [12]. Red ellipses indicate mound seep site regions with giant plumes. Red stars indicate gas hydrate outcrops.

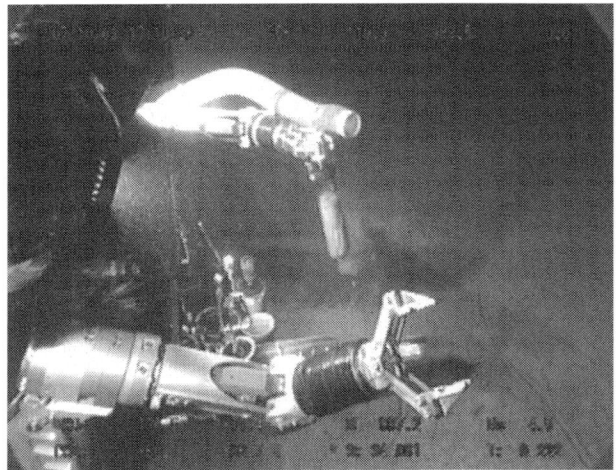

(a)

(b)

Figure 6: ROV Hyper Dolphin photos showing push coring operation at non-gas hydrate seep site (756-MB-1) for reference at Joetsu Knoll (a) and at gas hydrate mound (833-MBL-1) at Umitaka Spur (b). Note carbonate nodules on the seafloor at gas hydrate mound.

For TOC, TN, and $\delta^{13}C_{org}$ analysis, sediment samples were powdered and treated using 10% HCl solution to remove carbonate. An aliquot of each sample was preserved for analysis of TC, with no acid treatment, to calculate TIC (TC-TOC) and to control the quality of the acid treatment. Acidified samples were dried on a hot plate at 55°C for 1 day and later in an oven at the same temperature for 4 additional days. The weight of each dried sample was measured before and after acid treatment for later normalization, considering possible salt formation and weight increase [13, 14].

Ca. 20 mg of each sample were analyzed with a Thermo Finnigan Flash EA 1112 series CNS analyzer at the laboratory of the Department of Earth and Planetary Science of the University of Tokyo, using a retention time of 720 s. The analytical error was <0.2 wt% for TOC, <0.02 wt% for TN, and <0.1 wt% for TS, using sulfamethazine standard. The reproducibility error for duplicate analysis was <0.5 wt% for both TOC and TS, while TN had error <0.05 wt%.

A combination of Thermo Finnigan Flash EA 1112 series analyzers, CONFLO III, and Delta Plus mass spectrometer was used to analyze $^{13}C_{org}$. The weight of each sample was ca. 1.5 mg, depending on the TOC content. The analytical error, using standard IAEA-C6 sucrose, was <0.1% relative to Peedee belemnite (PDB) and the reproducibility error of duplicate analysis was <0.1%.

RESULTS

Lithology

Five lithologic units were identified from core descriptions from the bottom to the top (Figure 7). Unit 5 is at the bottom and is characterized by light gray bioturbated silty mud belonging to the early LGM sediments [13,14]. Unit 4 (TL-2) is characterized by thinly laminated dark gray mud deposited under anoxic conditions during the LGM lowstand.

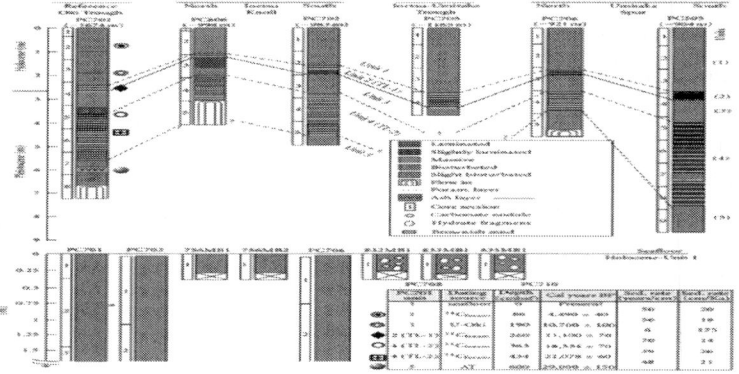

Figure 7: Piston-core correlation between background areas and mound-seep sites at both Umitaka Spur and Joetsu Knoll. Sedimentation rates for PC701 (Oki Trough) based on tephrostratigraphy and ^{14}C of foraminifera are shown. In the bottom, a correlation is shown between the shallowest parts of piston cores and push cores collected over the mounds. This correlation was modified from Freire [13, 14].

Unit 4 represents the TL-2 described by Tada et al. [20] and others, one of the most important and widespread layers of the late Quaternary in the Japan Sea. Unit 3 is a slightly bioturbated light gray silty mud and represents the LGM/Holocene transition [13, 14]. Unit 2 (TL-1) is a 5–20 cm dark gray thinly laminated mud layer, which is also common in the Japan Sea. TL-1 was deposited under anoxic/suboxic bottom water conditions as a result of the water stratification caused by the inflow of fresh water during the beginning of the Holocene [20]. Finally, Unit 1 is characterized by light gray bioturbated mud and represents the oxic bottom water conditions from the early-middle Holocene to the present [13, 14].

Tephra were identified in PC701 and their glass shards and composition were correlated [13, 17] with the Atlas of Tephra in and around Japan [21]. The upper tephra is a pumice type in Unit 1, ca. 50 cm above the top of Unit 2 (TL-1) at 1.88 m below the sea floor (mbsf-Figure 7). It was identified as Ulreung-Oki (U-Oki) tephra (10.7 ka cal BP) [21]. The lower tephra is a bubble wall glass type and is in the Unit 5, ca. 50 cm below the base of Unit 4 (TL-2) at 5.95 mbsf (Figure 7). Both the shape and the composition of the shards are well correlated with the Aira-Tanzawa (AT) tephra (28-29 ka cal BP) [21].

A total of four foraminifera samples were collected from PC701 for ^{14}C dating: one sample ofNeogloboquadrina dutertrei (warm water planktonic) at 0.80 mbsf in Unit 1 and three of Globigerina umbilicata (cold water planktonic) at 2.60 mbsf, in Unit 2 (TL-1) and at 3.63 and 4.34 mbsf respectively, both in Unit 4 (TL-2). Sedimentation rates were estimated for PC701 based on both tephra and foraminiferal data (Table 1) and inferred for the cores located in Joetsu Basin (Figure 7).

Table 1:

PC701 (unit)	Dating source	Depth (mm)	Age (cal ka BP)	Sedimentation rates (mm/ka)
1	Estimate	0	Present?	177
1	$^{14}C_{foram}$	800	4,490 ± 40	177
1	U-Oki	1900	10,700 ± 100	1750
2 (TL-1)	$^{14}C_{foram}$	2600	11,100 ± 70	133
4 (TL-2)	$^{14}C_{foram}$	3630	18,334 ± 70	254
4 (TL-2)	$^{14}C_{foram}$	4340	21,078 ± 60	221
5	AT	6000	28-29	

A depth-age conversion was made on the basis of tephra, ^{14}C of foraminifera, and lithologic correlation [13,14, 17]. Consequently, it was possible to obtain a good correlation between the reference core PC701 and those from the Joetsu Basin (Figure 7). The top of Unit 2 (TL-1) is inferred to occur at around 11.0 ka cal BP, while the top of Unit 3 is inferred to occur at 12.5 ka cal BP. The top of Unit 4 (TL-2) is placed around 18.0 ka cal BP and the bottom at 26.0 ka cal BP, which represents the top of Unit 5. The top of Unit 1 is the seafloor. High sedimentation rates were observed in PC701 (Figure 7), located at a favorable depositional site (Oki Trough depocenter). Samples below AT were estimated by extrapolation.

TOC, $\delta^{13}C_{org}$, TOC/TN, and TS/TOC Signatures

Graphs comparing age versus TOC (wt %), age versus $\delta^{13}C_{org}$ (‰ PDB), age versus TOC/TN, and age versus TS/TOC were constructed to promote an age-based correlation between Joetsu Basin and Oki Trough (Figure8). TOC varies from 0.5 to 2.0 wt% and $\delta^{13}C_{org}$ varies from −22.5 to −26.0% in Unit 5 at both locations (open sea and enclosed bay). TOC/TN strongly oscillates from 9 to 58, while TS/TOC ranges from near zero to 2.5. Unit 4 (TL-2) is characterized by TOC ca. 0.5 to 1.5 wt% and $\delta^{13}C_{org}$ varying from −23.0 to −26.0‰. TOC/TN strongly oscillates from 14 to 96 and TS/TOC varies from 0.5 to 2.0 (Figure 8).

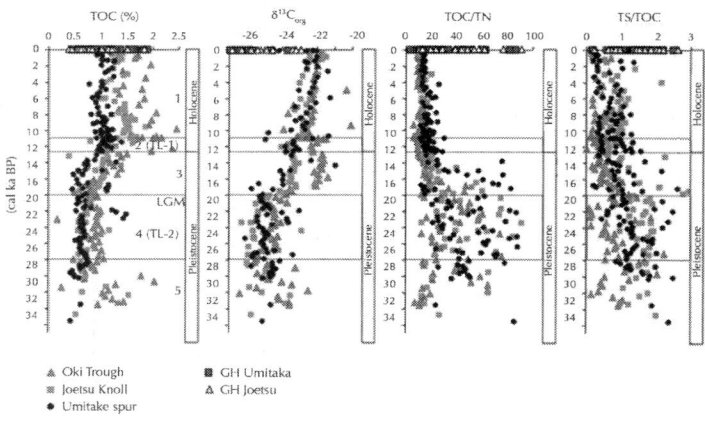

Figure 8: Correlation of cores from background area (red diamonds) and mounds of both Umitaka Spur (black circles) and Joetsu Knoll (green squares). The graphs show the geochemical parameters versus age, obtained by core correlation. Geochemical signatures of sediments placed over mounds, associated to gas hydrates, are a mixture of both Holocene and LGM signatures, in spite of being located at the seafloor and where only Holocene sediments expected to occur. Purple squares and open diamonds represent samples collected from both piston and push cores located at mounds of Umitaka Spur, where gas hydrate occurs. They are plotted at the top in the absence of age data for these cores. Their signatures range from the LGM to the present, indicating a mixture on the surface of mounds.

TOC increases from 0.5 to 1.5 wt% and $\delta^{13}C_{org}$ varies from −23.8% at the base to −21.7% at the top of Unit 3, representing the

LGM/Holocene transition. TOC/TN decreases from 46 at the base of Unit 3 to around 10 at the top, while TS/TOC oscillates from 0.5 to 2.5, mainly in cores located at Joetsu Basin.

Unit 2 (TL-1) is characterized by TOC from 1.0 to 2.5 wt%, and $\delta^{13}C_{org}$ varying from −21.0 to −24.0‰. TOC/TN is <20 and TS/TOC oscillates from near zero to 1.0. Unit 1 is characterized by TOC with a maximum value of 2.5 wt% at the base to a minimum of around 1.0 wt% at around 6.5 ka cal BP at open sea conditions (Figure 8). The $\delta^{13}C_{org}$ varies from −22.4% at the base to −20.1% in the upper part, representing the present Holocene sedimentation pattern in the area. TOC/TN oscillates nearly 10 on average and TS/TOC ranges from near zero to 1.0 (Figure 8).

Gas hydrate fragments and carbonate nodules recovered from mounds obstructed the penetration of piston cores, causing low recovery of sediments. All the piston-cores in these areas recovered surface sediments shallower than 2.0 mbsf despite the 4 to 6 m piston corer tube length used. In the same way, push cores recovered the shallowest 50 cm of seafloor sediments.

At these sites, the lithologic correlation is inaccurate due to the disturbance of sediments during coring operation. However, it is reasonable to assume that coring recovered shallowest sediments present at the seafloor (Figure 9).

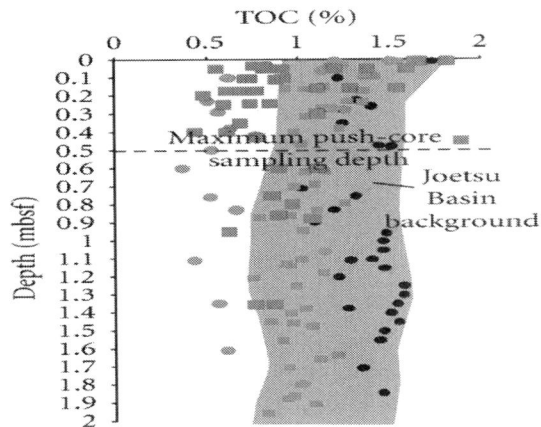

(a)

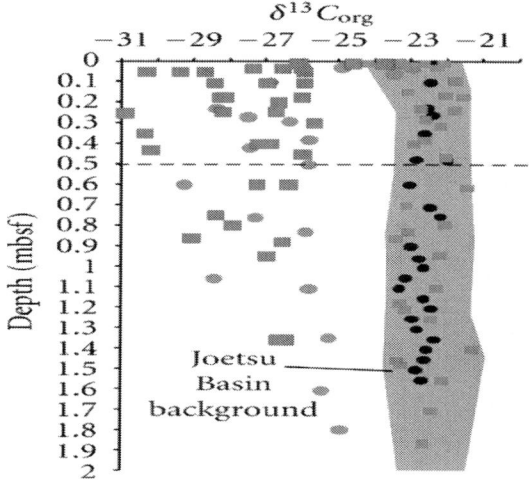

(b)

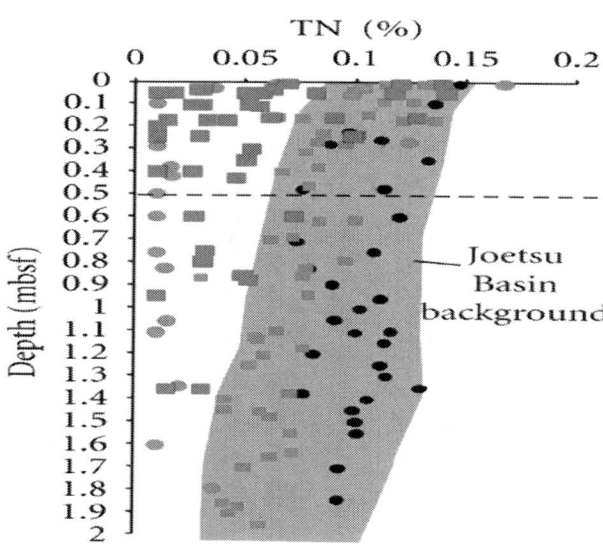

(c)

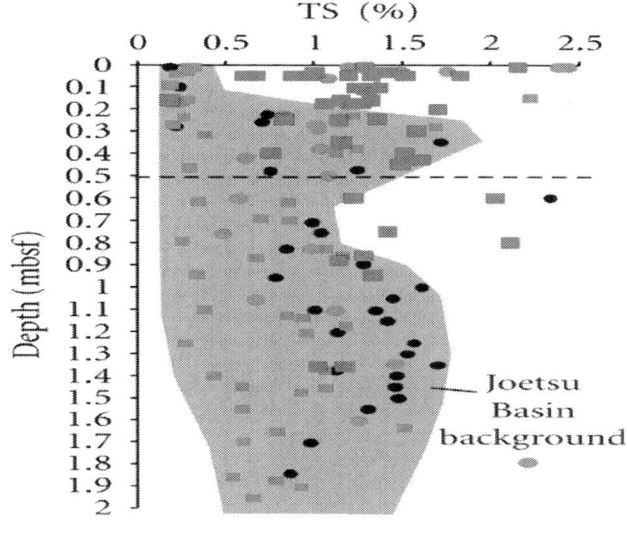

(d)

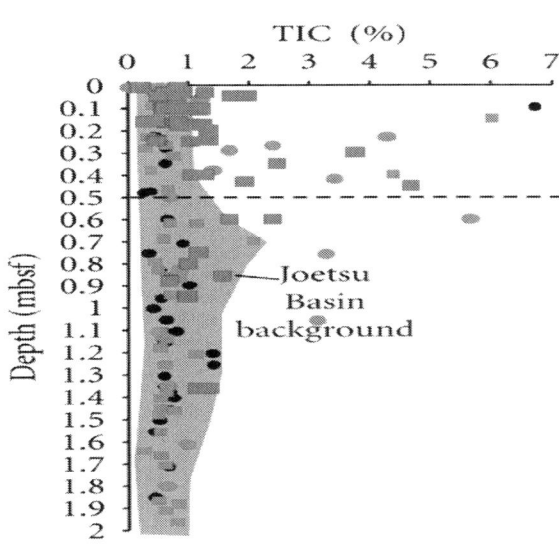

(e)

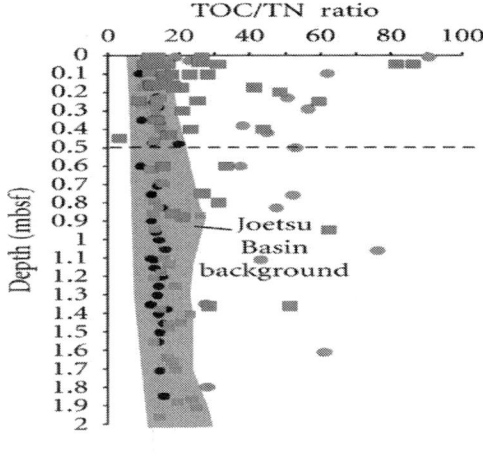

(f)

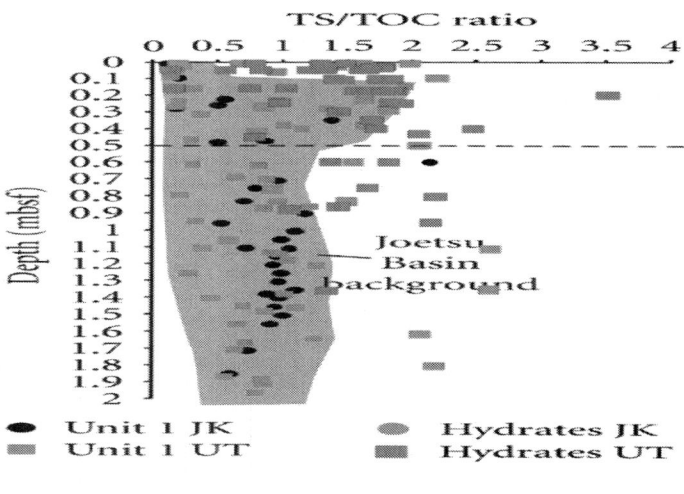

(g)

Figure 9: Piston and push core correlations from background areas and mounds of both Umitaka Spur and Joetsu Knoll. The graphs show the geochemical parameters versus depth. Only the first 2.0 mbsf were plotted.

Geochemical signatures of sediments placed over mounds, associated with gas hydrates, reflect a mixture of both Holocene and LGM signatures. Black circles indicate samples collected at Joetsu Knoll with no gas hydrates. Green squares represent samples collected at Umitaka Spur with no gas hydrates. Blue circles indicate gas hydrate-bearing sediments collected at mounds of Joetsu Knoll, while red squares represent gas hydrate-bearing sediments collected at mounds of Umitaka Spur. Shadow areas suggest geochemical trends of Unit 1 (Holocene).

DISCUSSION

The shallowest Holocene sediments of Joetsu Basin, until ca. 2.0 mbsf, are characterized by high TOC and TN contents, followed by low TOC/TN values (Figure 9). Enriched $\delta^{13}C_{org}$ signatures, combined with low TOC/TN values, indicate a predominance of marine organic matter deposited under oxic conditions, suggested by low TS/TOC values [13, 14]. On the other hand, the LGM sediments are characterized by low TOC and TN values, depleted $\delta^{13}C_{org}$ signatures, and high TOC/TN values, indicating a predominance of C_3-terrestrial organic matter deposited under anoxic conditions, suggested by high TS/TOC values [13, 14]. These values reflect that the terrestrial organic matter input was strongly controlled by the sea level changes that occurred since the LGM [14].

At mounds, however, shallow sediments from depths equivalent to those of the surrounding background Holocene sediments are strongly depleted in $\delta^{13}C_{org}$ followed by high TOC/TN values, similar to those of the underlying LGM sediments (Figure 9).

Elevated TS/TOC values are observed in the near seafloor sediments at mounds. These values can be explained by the anaerobic oxidation of methane (AOM) caused by the shallow boundary between the sulfate reduction and the methanogenesis zones.

AOM occurs within the ocean floor sediments, where sea waters, enriched in sulfate, meet methane formed at both methanogenesis and/or thermogenesis zones. This interface is named sulfate-methane interface (SMI) or transition (SMT) [22]. At the SMT the precipitation of sulfide and carbonate is common, which can explain the higher TS content and the presence of authigenic carbonate nodules in shallow sediments of mounds. AOM can be represented as the following equation:

$$CH_4 + SO_4^- \longrightarrow HCO_3^- + HS^- + H_2O$$

(1)

However, this reaction cannot explain the anomalous depleted values of $^{13}C_{org}$ (−26‰ to −31‰) and high TOC/TN signatures (almost >20), in an opposite trend of the neighboring Holocene sediments. Based on the age-based graphs (Figure 8) it is possible to infer that the gas hydrate-bearing sediments, located at the top of mounds, have similar geochemical signatures of those of the LGM sediments. On the other hand, samples collected in the mound flanks are apparently similar to the Holocene samples of Unit 1. Gas hydrate-bearing sediments are strongly disturbed but, in spite of this, Dr. Fumio Akiba (personal communication) was able to identify diatom species from the LGM in those sediments located on the top of mounds (Figure 10). On the other hand, he found Holocene species at similar depths in cores located in the flanks of the same mounds.

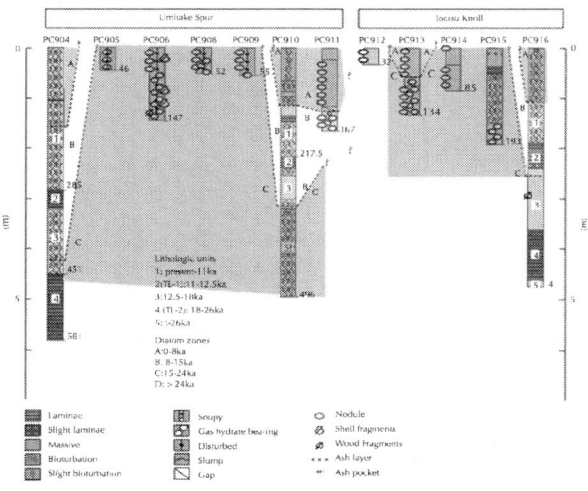

Figure 10: Correlation between cores located in the flanks of mounds and cores located in the top of mounds of both Umitaka Spur and Joetsu Knoll. Diatom zonation was provided by Dr. Fumio Akiba (personal communication). See Figure 1 for core locations.

The diatom analysis of eight piston cores has revealed that they contained common to abundant diatom assemblages, and they can be clearly subdivided into four diatom zones as A, B, C, and D zones, in descending order, primarily based on the limited occurrences of two marker diatoms in the studied sediments:Fragilariopsis dolilus and Thalassiosira hyperborea. The former is a warm water species, and it is common to abundant in A Zone. The latter is a cold water species often associated with ice sheets and low-salinity water, and rare to common occurrence is recognized in C Zone. The B Zone is an interval between the A and C zones, and the D Zone is a horizon below the C Zone (Figure 10).

The characteristic occurrences of two diatoms in the late Quaternary sediments of the Japan Sea have been well noticed previously and were linked with dated paleoceanographic events of Japan Sea [23–25]. The diatom zones A, B, C, and D range from 0–8 ka, 8–15 ka, 15–24 ka, and >24 ka, respectively.

Crossplots of geochemical parameters are strong tools to evaluate the characteristics of both Holocene and LGM sediments. The crossplot TOC/TN versus TS/TOC (Figure 11(a)) illustrates that the sediments of Unit 1 and almost all of the sediments from Unit 2 (TL-1), both from the Holocene, plot in the field corresponding to marine organic matter deposited under oxic conditions. In the same graph it is possible to observe that sediments from Unit 3 (LGM-Holocene transition) partially plot in both marine/oxic and terrestrial/anoxic fields, while almost all samples from both LGM Unit 4 (TL-2) and Unit 5 plot in the terrestrial/anoxic field (Figure 11(a)). Gas hydrate-bearing sediments recovered from the top of mounds, however, have signatures partially compatible to those of LGM sediments, suggesting a mixture with Holocene sediments.

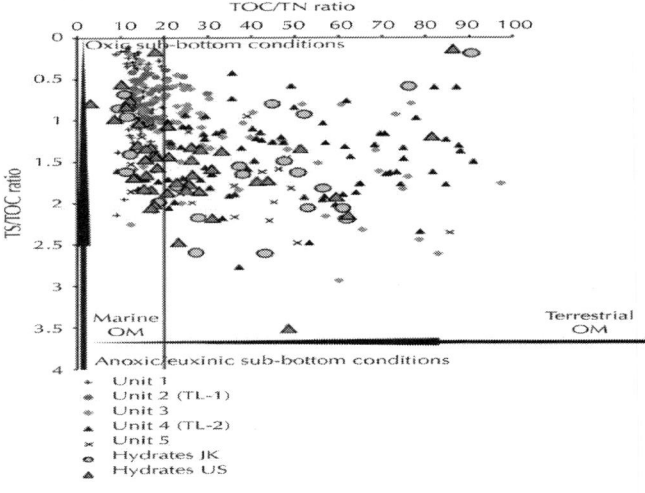

(a)

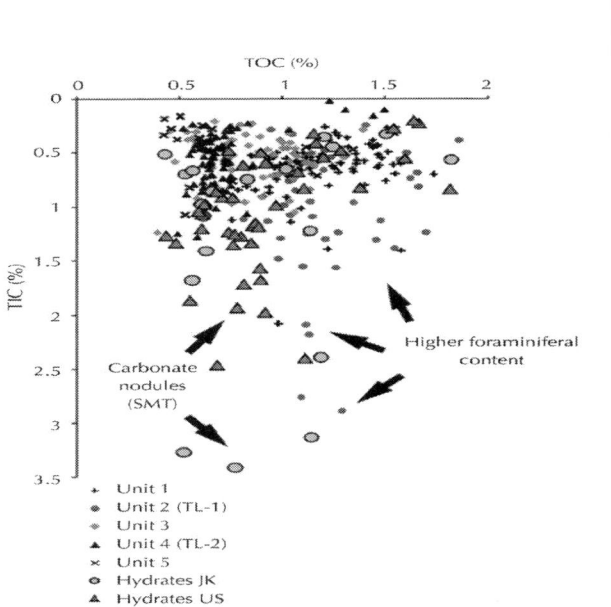

(b)

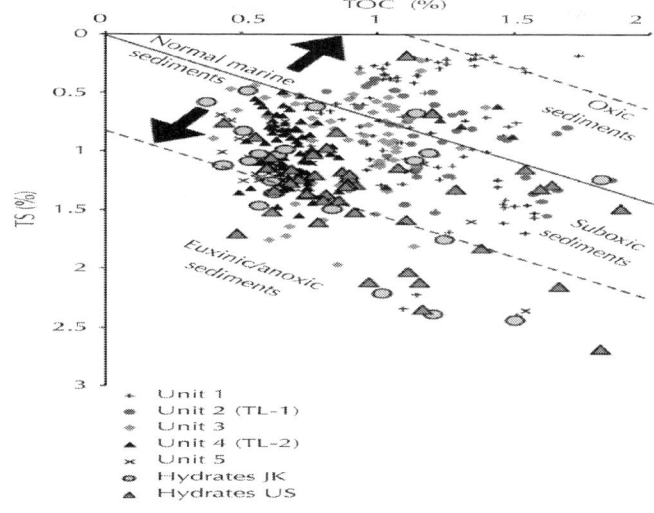

(c)

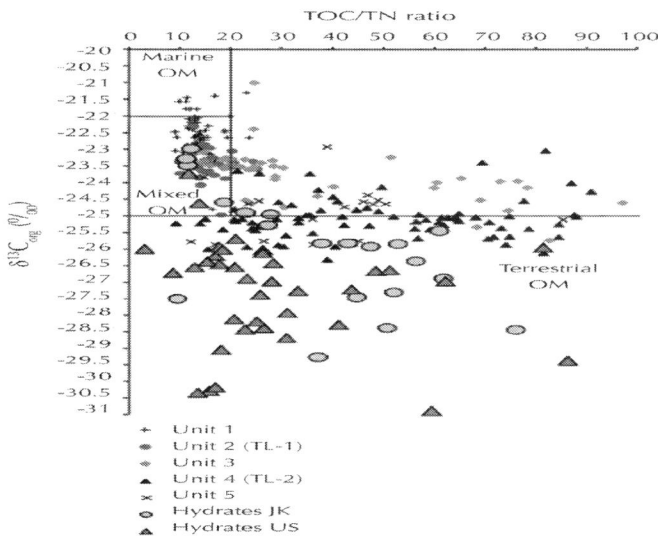

(d)

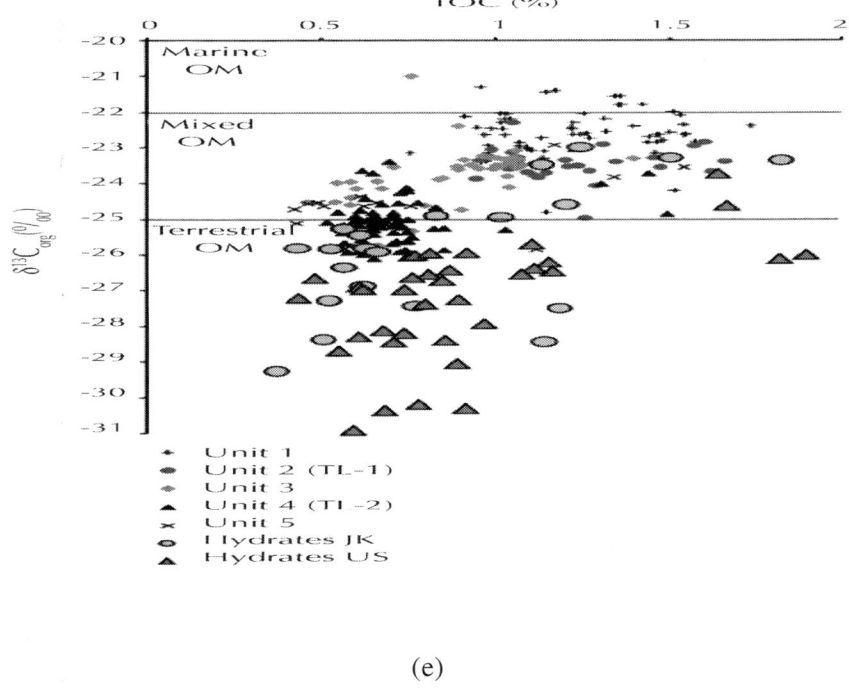

(e)

Figure 11: Crossplots of TOC/TN versus TS/TOC (a), TOC versus TIC (b), TOC versus TS (c), TOC/TN versus (d) and TOC versus . All parameters suggest a mixture of both Holocene and LGM geochemical signatures on the top of mounds, indicating an uplift of deeper sediments related to the formation of mounds.

Excepted when related to foraminifera-enriched layers [14], the crossplot TOC versus TIC (Figure 11(b)) shows that Holocene sediments from both Unit 1 and Unit 2 (TL-1) have high TOC content and relatively low TIC content. However, superficial gas hydrate-bearing sediments are strongly enriched in TIC caused by authigenic carbonate precipitation as a result of AOM.

The crossplot TOC versus TS (Figure 11(c)) is useful to differentiate euxinic from normal marine (noneuxinic) sediments [20]. Sediments of Unit 1 and Unit 2 (TL-1) partially plot on both oxic- and suboxic type fields. Gas hydrate-bearing sediments, however, plot totally in both suboxic-type and euxinic-type fields, similar to those sediments of Unit 4 (TL-2).

The crossplot TOC/TN ratio versus $\delta^{13}C_{org}$ (Figure 11(d)) suggests that geochemical signatures of gas hydrate-bearing sediments are partially similar to those of LGM sediments, suggesting the mixture of different origins. Such similarities are also confirmed by the graph TOC versus $^{13}C_{org}$ (Figure 11(e)). This graph shows that gas hydrate-bearing sediments are strongly depleted in $\delta^{13}C_{org}$ in a similar trend of the LGM sediments of Unit 4 (TL-2).

Based on the geochemical signatures (Figure 11), supported by the disappearance of the Holocene diatom zones A and B in the shallow gas hydrate-bearing sediments at the top of the mounds (Figure 10), the following process is proposed here to explain these unusual characteristics.

- The formation of gas hydrate nodules and blocks (Figure 3) causes an increase in the pore space of sediments located around the fault-conduits and within the fault planes themselves.

- The increase of volume in both pore space and fault planes causes an increase in the volume of the sediments as a whole. Combined with this, an intense gas/water pore pressure toward the top induces a strong upward movement promoting the uplift of the sediments and the formation of the mounds at the seafloor. The pore pressure is created by the upward migration of both methane and water released from deeper gas hydrate dissociation at the base of the GHSZ, or direct from deeper hydrocarbon reservoirs below the BSR [11, 13].

- On the seafloor, erosion or nondeposition of the Holocene sediments at the top of mounds can explain the absence of the Holocene sediments and the consequent exposure of deeper LGM sediments.

Another proof of the proposed uplift of sediments is the presence of carbonate nodules at the seafloor over mounds. Methane-related carbonate nodules were not formed at the seafloor. They were formed by AOM at the SMT, below the seafloor [22].

Carbonate nodules recovered in piston cores from mounds of both Joetsu Knoll and Umitaka Spur show ^{14}C ages ranging from 26 cal ka BP to 47 cal ka BP at depths randomly varying from 0 to around 4 mbsf, indicating that they were formed during the LGM [26], in spite of the nodules, older than 35 cal ka, and should not be considered due to the potential presence of dead carbon.

On the other hand, carbonate nodules were observed in some Holocene sediment (Figure 11), indicating recent generation of authigenic minerals caused by active methane seep and gas hydrate dissociation.

This may indicates that carbonate nodules formed at different times are now exposed on the seafloor, and they need to be uplifted first, and then eroded to be spread out around the mounds (Figure 6). On the other hand, only erosion cannot explain the phenomena observed at both Umitaka Spur and Joetsu Knoll because all the seep sites are located at mound areas, not favorable for sediment deposition (Figure 6).

The combination of all these factors indicate that the uplift of pre-Holocene sediments to the surface of the seafloor, followed by the erosion or nondeposition of the Holocene sediments, is the better explanation for the growth of mounds of both Umitaka Spur and Joetsu Knoll.

The uplift >2 m, inferred by correlation using both geochemical and sedimentological parameters, was caused by the increase in the sedimentary pore space and the consequent increase of the volume of the whole sedimentary section creating sediment uplift. This process is still ongoing at the present time, and the gas/water migration from deeper zones through faults and fractures amplifies the uplift and creates the gas seeps and plumes observed over mounds.

A model of the uplift of solid materials caused by the increase in the volume of the sedimentary pore space, associated with a migration of gas and water released after gas hydrate dissociation is presented in Figure 12.

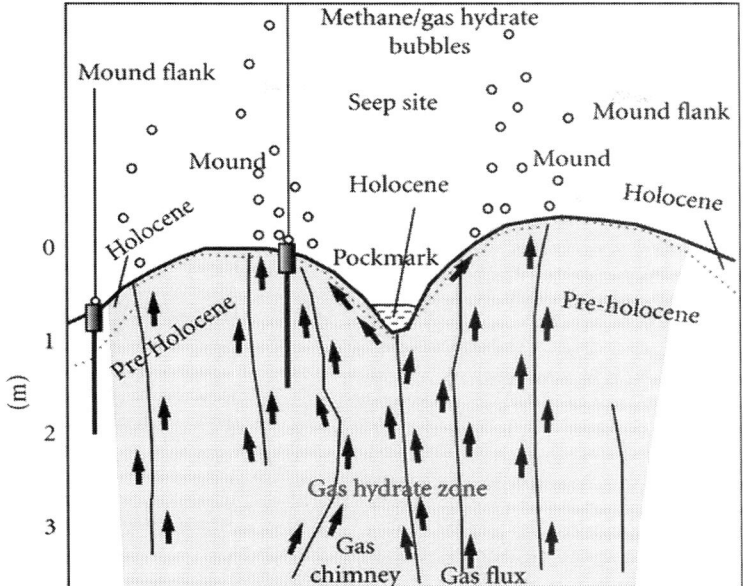

Figure 12: Migration model for the formation of mounds. During gas hydrate dissociation and direct gas migration along faults from deeper reservoirs, LGM or older sediments are uplifted and exposed on the seafloor. The exposure of old sediments is caused by an increase in the pore-space volume that induces an uplift of pre-Holocene sediments resulting in seafloor mounds. This process is still ongoing at the present time.

SUMMARY

Based on this investigation, it is possible to conclude that the formation of gas hydrates and authigenic carbonate, associated with the pore-pressure growth caused by gas hydrate dissociation, promotes an increase in the volume of pore space inducing the formation of mounds at the seafloor. The uplift of older sediments is followed by nondeposition or erosion of the Holocene sediments on the top of the mounds.

ACKNOWLEDGMENTS

The authors are thankful to R. O. Kowsmann for comments. Thanks go to the editor and anonymous reviewers for their comments and suggestions.

REFERENCES

1. K. Tamaki and N. Isezaki, "Tectonic synthesis of the Japan Sea based on the collaboration of the Japan-URSS Monograph Project," in Geology and Geophysics of the Japan Sea, N. Isezaki, et al., Ed., vol. 1 of Japan-URSS Monograph Series, pp. 483–487, 1996.

2. L. Jolivet, K. Tamaki, and M. Fournier, "Japan Sea, opening history and mechanism: a synthesis,"Journal of Geophysical Research, vol. 99, no. 11, pp. 22,237–22,259, 1994.

3. A. Takeuchi, "Recent crustal movements and strains along the eastern margin of Japan Sea floor," inGeology and Geophysics of the Japan Sea, N. Isezaki, et al., Ed., vol. 1 of Japan-URSS Monograph Series, pp. 385–398, 1996.

4. K. Tamaki, "Geological structure of the Japan Sea and its tectonic implications," Bulletin, vol. 39, pp. 269–365, 1988.

5. A. Okui, M. Kaneko, S. Nakanishi, N. Monzawa, and H. Yamamoto, "An integrated approach to understanding the petroleum system of a frontier deep-water area, offshore Japan," Petroleum Geoscience, vol. 14, no. 3, pp. 223–233, 2008.

6. K. Tamaki and E. Honza, "Incipient subduction and deduction along the eastern margin of the Japan Sea," Tectonophysics, vol. 119, no. 1–4, pp. 381–406, 1985.

7. U. Suzuki, "Petroleum geology of the Sea of Japan, Northern Honshu," Journal of the Japanese Association For Petroleum Technology, vol. 44, no. 5, pp. 291–307, 1979 (Japanese).

8. T. Seno, "Syntheses of the regional stress fields of the Japanese islands," The Island Arc, vol. 8, no. 1, pp. 66–79, 1999.

9. A. Taira, "Tectonic evolution of the Japanese island arc system," Annual Review of Earth and Planetary Sciences, vol. 29, pp. 109–134, 2001.

10. B. K. Son, T. Yoshimura, and H. Fukasawa, "Diagenesis of dioctahedral and trioctahedral smectites from alternating beds in miocene to pleistocene rocks of the Niigata Basin, Japan," Clays and Clay Minerals, vol. 49, no. 4, pp. 333–346, 2001.

11. A. F. M. Freire, R. Matsumoto, and L. A. Santos, "Structural-stratigraphic control on the Umitaka Spur gas hydrates of Joetsu Basin in the eastern margin of Japan Sea," Marine and Petroleum Geology, vol. 28, no. 10, pp. 1967–1978, 2011.

12. R. Matsumoto, "Formation and collapse of gas hydrate deposits in high methane flux area of the Joetsu Basin, eastern margin of Japan Sea," Journal of Geography, vol. 118, no. 1, pp. 43–71, 2009 (Japanese), In R. Matsumoto, Ed., Special issue on "Methane hydrate (part1): occurrence, origin, and environmental impact".

13. A. F. M. Freire, An integrated study on the gas hydrate area of Joetsu Basin, eastern margin of Japan Sea, using geophysical, geological and geochemical data [Ph.D. thesis], The University of Tokyo, Graduate School of Frontier Sciences, 2010.

14. A. F. M. Freire, T. R. Menezes, R. Matsumoto, T. Sugai, and D. J. Miller, "Origin of organic matter in the Late-Quaternary sediments of the eastern margin of Japan Sea," Journal of the Sedimentological Society of Japan, vol. 68, pp. 117–128, 2009.

15. R. Matsumoto, "Methane plumes over a marine gas hydrate system in the eastern margin of the Japan Sea: a possible mechanism for the transportation of subsurface methane to shallow waters," inProceedings of the 5th International Conference on Gas Hydrates, pp. 749–754, Trondheim, Norway, 2005.

16. T. Oba, M. Kato, H. Kitazato, et al., "Paleoenvironmental changes in the Japan Sea during the last 85,000 years," Paleoceanography, vol. 6, no. 4, pp. 499–518, 1991.

17. A. F. M. Freire, T. Sugai, and R. Matsumoto, "The use of tephras for stratigraphic correlation: a case study on the eastern margin of Japan Sea," Boletim de Geociências da Petrobras, vol. 18, pp. 97–121, 2010 (Portuguese).

18. A. Hiruta, G. T. Snyder, H. Tomaru, and R. Matsumoto, "Geochemical constraints for the formation and dissociation of gas hydrate in an area of high methane flux, eastern margin of the Japan Sea," Earth and Planetary Science Letters, vol. 279, no. 3-4, pp. 326–339, 2009.

19. H. Tomaru, Z. Lu, G. T. Snyder, U. Fehn, A. Hiruta, and R. Matsumoto, "Origin and age of pore waters in an actively venting gas hydrate field near Sado Island, Japan Sea: interpretation of halogen and [129]I distributions," Chemical Geology, vol. 236, no. 3-4, pp. 350–366, 2007.

20. R. Tada, T. Irino, and I. Koizumi, "Land-ocean linkages over orbital and millennial timescales recorded in late Quaternary sediments of the Japan Sea," Paleoceanography, vol. 14, no. 2, pp. 236–247, 1999.

21. H. Machida and F. Arai, Atlas of Tephra in and Around Japan, University of Tokyo press, Tokyo, Japan, 2003.

22. G. R. Dickens, "Sulfate profiles and barium fronts in sediment on the Blake Ridge: present and past methane fluxes through a large as hydrate reservoir," Geochimica et Cosmochimica Acta, vol. 65, no. 4, pp. 529–543, 2001.

23. I. Koizumi, "The Japan Sea after last glacial stage: Diatoms –with special reference to the analysis of KH-79-3, C-3 core," Monthly-Chikyu, vol. 6, pp. 547–552, 1984 (Japanese).

24. I. Koizumi, "Last glacial sediments from the seafloors around Japan," Monthly-Kaiyo, vol. 7, pp. 338–343, 1985 (Japanese).

25. I. Koizumi, the Japan Sea and Circum-Japan Sea Areas—Their Establishment and Change of Natural Environments, Katokawa-gakugei-shuppan, Tokyo, Japan, 2006.

26. Y. Watanabe, S. Nakai, A. Hiruta, R. Matsumoto, and K. Yoshida, "U-Th dating of carbonate nodules from methane seeps off Joetsu, eastern margin of Japan Sea," Earth and Planetary Science Letters, vol. 272, no. 1-2, pp. 89–96, 2008.

Chapter 4

Exploring and Using the Magnetic Methods

Othniel K. Likkason[1]

[1]Physics Programme, Abubakar Tafawa Balewa University, Bauchi, Nigeria

INTRODUCTION

The Earth is principally made up of three parts: core, mantle and crust (Fig. 1). As understood today, right at the heart of the Earth is a solid inner core composed primarily of iron. At 5, 700°C, this iron is as hot as the Sun's surface, but the crushing pressure caused by gravity prevents it from becoming liquid. Surrounding this is the outer core, a nearly 2, 000 km thick layer of iron, nickel, and small quantities of other metals. Lower pressure than the inner core means the metal here is fluid. Differences in temperature, pressure and composition within the outer core cause convection currents in the molten metal as cool, dense matter sinks while warm, less dense matter rises. This flow of liquid iron generates electric currents, which in turn produce magnetic fields (Earth's field). These convection processes in the liquid part of core (outer core) give rise to a dipolar geomagnetic field that resembles that of a large bar magnet aligned approximately along the Earth's rotational axis. The mantle plays little part in the Earth's magnetism, while interaction of the past and present geomagnetic field with the rocks of the crust produces magnetic anomalies recorded in detailed when surveys are carried out on or above the Earth's surface.

The magnitude of the Earth's magnetic field averages to about 5×10^{-5} T (50, 000 nT). Magnetic anomalies as small as 0.1 nT can be measured in continental magnetic surveys and may be of geological significance.

The magnetic methods, perhaps the oldest of geophysical exploration techniques bloomed after the World War II. Today, with improvements in instrumentation, navigation and platform compensation, it is possible to map the entire crustal section at a variety of scales from strongly magnetic basement at a very large scale to weakly magnetic sedimentary contacts at small scale. Methods of magnetic data treatment, filtering, display and interpretation have also advanced especially with the advent of high performance computers and colour raster graphics.

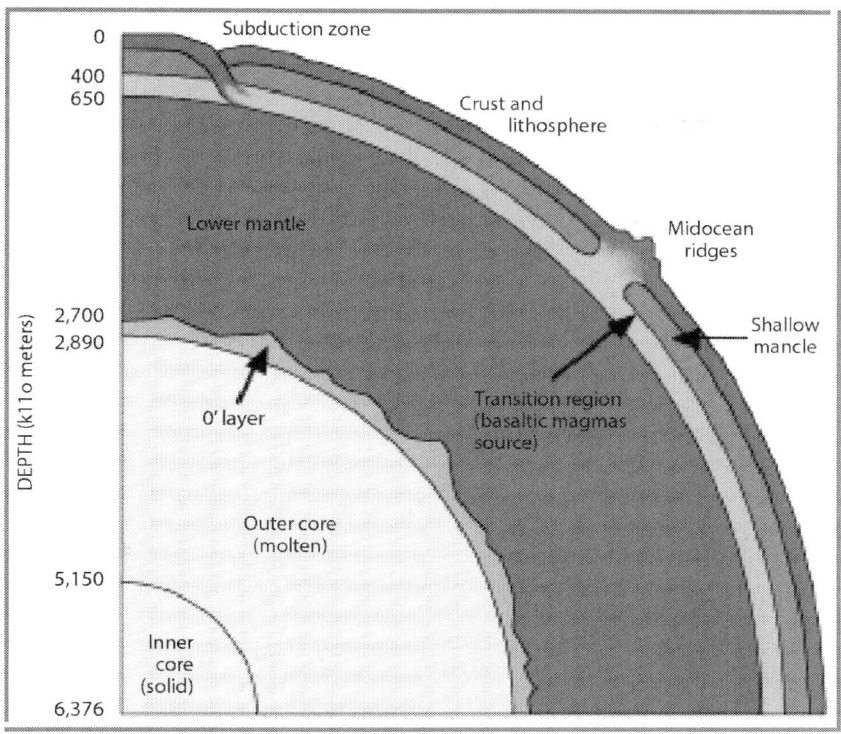

Figure 1: Internal structure of the Earth (from http://zebu.uoregon.edu).

As is well known today, magnetic methods are used to solve various problems such as:

- Mapping the basement surface and sediments in oil/gas exploration
- Detecting different types of ore bodies in mining prospecting
- Detecting metal objects in engineering geophysics
- Mapping basement faults and fractures
- Determining zones with different mineralization in logging as well as inspecting casing parameters
- Studying the magnetic field of the Earth and its generators and
- A variety of other purposes such as natural hazards assessment, mapping impact structures and environmental studies.

Magnetic observations are obtained relatively easily and cheaply and a few corrections are applied to them. This explains why the magnetic methods are one of the most commonly used geophysical tools. Despite these obvious advantages, interpretations of magnetic observations suffer from a lack of uniqueness due to dipolar nature of the field and other various polarization effects. Geologic constraints, however, can considerably reduce the level of ambiguity. Information from magnetic surveys comes from rock units at depth as well as from those at or near the surface. This is the strength of the magnetic method (or any surface geophysical method), making it more powerful than any other remote sensing method which relies on the information from reflections of electromagnetic (EM) waves by materials on the Earth surface. Thus while the natural magnetic field of the Earth is measured in magnetic method, EM radiation normally is used as the information carrier in remote sensing. Electromagnetic radiation is a form of energy with the properties of a wave, and its major source is the sun. Solar energy traveling in the form of waves at the speed of light is known as the electromagnetic spectrum. Passive remote sensing systems record the reflected energy of electromagnetic radiation or the emitted energy from the Earth, such as cameras and thermal infrared detectors. Active remote sensing systems send out their own energy and record the reflected portion of that energy from the Earth's surface, such as radar imaging systems.

In this chapter, we explore the magnetic methods of geophysical exploration. The first part of the chapter covers the fundamental concepts of magnetic force field, the Earth's magnetic field and its relationship with gravity field. The second part deals with the measurement procedures and treatment of the magnetic field data, while the third part covers the magnetic effects of simple geometric bodies, processing and interpretation of magnetic data and ending it with treatment, analysis and interpretation of real field data.

FUNDAMENTAL MAGNETIC THEORIES

Any magnetic grain is a dipole. That is, it has two poles, P_1 and P_2 of opposite signs diametrically linked. Charles Augustin de Coulomb in 1785 showed that the force of attraction or repulsion between electrically charged bodies and between magnetic poles obeys an inverse square law similar to that derived for gravity by Newton.

The mathematical expression for the magnetic force, F_m experienced between two magnetic monopoles is given by:

$$F_m = \frac{1}{\mu} \frac{P_1 P_2}{r^2}$$

(1)

Where μ is a constant of proportionality known as the magnetic permeability, P_1 and P_2 are the 'strengths' of the magnetic monopoles and r is the distance between the poles.

We note that the expression in equation (1) is identical to the gravitational force, $F_g = G\frac{m_1 m_2}{r^2}$ and electrical force, $F_g = k\frac{q_1 q_2}{r^2}$ expressions. Here m_1, m_2 and q_1, q_2 are respectively masses and electrical charges separated by distance r, G is the universal gravitational constant while k is the Coulomb's law constant for the medium. However, unlike the gravitational constant, G, the magnetic permeability, μ is a property of the material medium in which the two monopoles, P_1 and P_2 are situated. If they are placed in a vacuum, then μ is for the free space. Also, unlike m_1 and m_2, P_1 and P_2 can be either positive or negative in sign. If P_1 and P_2 have the same sign, the force, F_m between the two monopoles is repulsive. If P_1 and P_2 have opposite signs, F_m is attractive.

We may seem to easily compare the gravitational force between masses m_1 and m_2 separated by r to that of either the attractive or repulsive magnetic force between two monopoles. However, the magnetic monopoles have never existed! Rather the fundamental magnetic element appears to consist of two magnetic monopoles: one positive and the other negative, separated by a distance. Thus the fundamental magnetic element consisting of two monopoles is called a magnetic dipole. Every magnetic grain is therefore a dipole.

We can therefore determine the force produced by a dipole by considering a force produced by two monopoles. Since the dipole is simply two monopoles, each of strength P_1 and P_2, we expect that the force generated by a dipole is simply the force generated by one monopole added vectorially to the force generated by the second monopole. Consequently, the force distribution for a dipole is nothing more than the magnetic force distribution observed around a simple bar magnet. Thus a bar magnet can be thought as two magnetic monopoles separated by a length of the magnet. The magnetic force appears to originate out of the North Pole (N) of the magnet and to terminate at the South Pole (S) of the magnet. Some of the field lines pass through the material of the magnet (high concentration because of high μ), some pass through air (low concentration because of low μ). Notice that even in air; the poles have high density of field lines. Also, the lines radiate out from N (vertically outward) and radiate into S (vertically inward). Between the length of the bar in air, the magnetic field directions are variable, but with the middle of the bar having a near horizontal field direction. Again, the field strength and direction at any point around the bar magnet is a vector sum of the force field contributed by each of the monopole (N or S).

When we examine equation (1) in terms of unit of measurement, we see that the magnetic force, F_m retains its fundamental unit of newton (N) and r^2 would be in square metre (m^2). Permeability, μ by the S. I. unit definition, is a unitless constant. The units of the pole strength, P are defined such that if a force of 1 N is produced by two unit poles separated by a distance of 1 m, then each unit pole has a strength of one ampere-metre (1 Am). Thus a unit pole has an S.I unit of ampere-metre.

We can also define, from equation (1), the force per unit pole strength exerted by a magnetic monopole, P_1 or P_2. This is called magnetic field strength or magnetizing force, H. Thus

$$H = \frac{F_m}{P_2} = \frac{P_1}{\mu r^2}$$

(2)

Here again, given the units associated with force (N) and magnetic monopoles (Am), the unit associated with magnetic field strength, H are N/A-m and by definition, 1 N/A-m is referred to as a tesla (T): named after a Croatian inventor, Nikola Tesla. Thus 1 T = 1 N/Am. Indeed from equation (2), the unit of H can be expressed as Am/m^2 or Am^{-1} (ampere per metre). Thus 1 N/Am = 1 Am^{-1} = 1 T. Similarly, the unit of magnetic flux is weber (Wb) and magnetic flux per unit area is the magnetic strength we have been talking about. Thus the unit of magnetic strength can also be expressed in weber per square metre (Wb/m^2). Hence 1 Wb/m^2 = 1 T.

When describing the magnetic field of the Earth, it is common to use units of nanotesla (nT), where 1 nT = 10^{-9} T. The average strength of the Earth's magnetic field, H is about 50, 000 nT (ranges from 20, 000 to 70, 000 nT). A nanotesla has the value as the old unit of gamma (1 nT = 1 gamma).

When magnetic materials or rocks are placed within a field, T (a magnetizing force such as H given in equation (2)), the magnetic materials or rocks will produce their own magnetizations or polarizations. This phenomenon is called induced magnetization, magnetic polarization or magnetic induction. The strength of the magnetic field induced on the magnetic material due to the inducing field, T is called the intensity of magnetization or magnetic polarization, J_i; where

$$J_i = kT$$

(3)

The constant of proportionality, k is the magnetic susceptibility and is a unitless constant determined by the physical properties of the magnetic material. The susceptibility, k can either be positive or negative in values. Positive values imply that the field, J_i is in the same direction as the inducing field T. Negative k implies that the induced magnetic field is in the opposite direction as the inducing field. Details of the mechanisms of induced magnetization can be further obtained from [1].

In magnetic exploration method, the susceptibility is the fundamental material property whose spatial distribution, we attempt to determine. We see that magnetic susceptibility is analogous to density in gravity surveying. Unlike density, there is a large range of susceptibilities even within materials and rocks of the same type. This definitely will put limit to knowledge of rock type through susceptibility mapping of an area.

Magnetic susceptibility in SI unit is a dimensionless ratio having a magnitude much less than 1 for most rocks. Hence a typical susceptibility value may be expressed (as for example) k = 0.0064 SI. In the old c.g.s. system of electromagnetic units (emu), the numerical value of magnetic susceptibility for a given specimen is smaller by a factor of 4π than the SI value. Thus k (SI) = k (emu) x 4π. Hence for k = 0.0064 SI, k (emu) = k (SI)/4π = 0.00051 emu.

THE EARTH'S MAGNETIC FIELD

Nearly 90% of the Earth's magnetic field (geomagnetic field) looks like a magnetic field that would be generated from a dipolar magnetic source located at the centre of the Earth and nearly aligned with the Earth's rotational axis. This field is believed to originate from convection of liquid iron in the Earth's outer core [2] and is monitored and studied using global network of magnetic observatories and various satellite magnetic surveys. If this dipolar description of the Earth's field were complete, then the magnetic equator would nearly correspond to the Earth's geographic equator and the magnetic poles would also nearly correspond to the geographic poles. The strength of the Earth's field at the poles is about 60, 000 nT. This is called the Main Field of the Earth. This field changes slowly with time and is believed to go through a decay and collapse, followed by polar reversal on a time scale of the order of 100, 000 years [3], [4]. The construction of a global magnetic reversal timescale is of fundamental importance in deciphering Earth's history. For details on such discussion, [5] can be consulted.

The remaining 10% of the Earth's magnetic field cannot be explained in terms of simple dipolar sources. The larger component of this 10% of the Earth's field originates in iron-bearing rocks near the Earth's surface where temperatures are sufficiently low (i.e. less than the Curie temperature of the rocks). This region is confined to the upper 30 – 40 km of the crust and is the source of the crustal field which is made up of induced field on magnetically susceptible rocks and remanent magnetism of the rocks. The smaller portion of the 10% comes from the upper atmosphere (external source).

The external source field is believed to be produced by interactions of the Earth's ionosphere with the solar wind. Hence some temporal variations (usually variable over hours at tens of nT or occasionally variable over a few hours at hundreds of nT: the magnetic storm) are correlated to solar activity. The external component (except for magnetic storm phenomenon) is usually regular and are corrected/removed appropriately from field measurements in a process similar to drift correction in gravity surveys. Where magnetic storm is detected, survey is most often discontinued until after the phenomenon has passed.

The crustal field, its relation to the distribution of magnetic minerals within the crust, and the information this relation provides about exploration targets are the primary subjects of the magnetic method in exploration. In a magnetic survey, the magnetic induction, B whose magnitude is measured at a point is the vector sum of four field components:

- The Earth's main field which originates from dynamo action of conductive fluids in the Earth's deep interior [6];
- An induced field caused by magnetic induction in magnetically susceptible earth materials polarized by the main field [7];
- A field caused by remanent magnetism of earth materials [7]; and
- Other (usually) less significant fields caused by solar, atmospheric [8] and cultural influences

While we can handle the external features (source 4) component (like drift correction in gravity survey: for the solar/atmospheric sources) and divesting from such features or recognizing their transient effects and removing them (for cultural features), the main field is examined from complex models that have been developed and are available. Our intent here is to characterize the global magnetic field (main field) in order to isolate the magnetic field caused by crustal sources (sources 2 and 3).

Spherical harmonic analysis provides the means with which to determine from measurements of a potential field and its gradient on a sphere whether the sources of the field lie within the sphere or outside the sphere. Carl Friederich Gauss in 1838 was the first to describe the geomagnetic field in this way and concluded that the observed field at the Earth's surface originates entirely from within the Earth. However, we know today from satellite observations, space probes and vast accumulation of information from field measurements that a small part of the geomagnetic field originates from outside the Earth.

We consider a magnetic induction vector, $\vec{B}$ at a point on or above the Earth's surface and its potential, V, such that $\vec{B} = -\vec{\nabla}V$. If we assume a source-free space, V is harmonic and satisfies Laplace's equation:

$$\nabla^2 V = 0$$

(4)

Following [9], if no sources exist outside the sphere, then both V

and $\dfrac{\partial V}{\partial r}$ must vanish for r→∞ and hence:

$$V^{i} = a \sum_{n=0}^{\infty} \left(\frac{r}{a}\right)^{n+1} \sum_{m=0}^{n} \left(A_n^{mi} \cos m\varphi + B_n^{mi} \sin m\varphi\right) P_n^{m}(\theta) : r \geq a$$

(5)

On the other hand, if all sources lie outside the sphere, then V and

$\dfrac{\partial V}{\partial r}$ must be finite within the sphere and appropriately,

$$V^{e} = a \sum_{n=0}^{\infty} \left(\frac{r}{a}\right)^{n} \sum_{m=0}^{n} \left(A_n^{me} \cos m\varphi + B_n^{me} \sin m\varphi\right) P_n^{m}(\theta) : r \geq a$$

(6)

Where in both equations (5) and (6), the superscripts, i and e denote internal and external sources respectively, θ is the co-latitude (latitude = 90° - θ), $\square$ is longitude, r is the radial distance from the centre of the sphere, a is the radius of the sphere, and $P_n^{m}(\theta)$ is an associated Legendre polynomial of degree n and order m normalized according to the convention of Schmidt. The magnitude of the normalized Schmidt surface harmonics when squared and averaged over the sphere can be expressed as

$$\frac{1}{4\pi r^2} \int_0^{2\pi} \int_0^{\pi} P_n^{m}(\theta) \begin{Bmatrix} \cos m\varphi \\ \sin m\varphi \end{Bmatrix} P_{n'}^{m'}(\theta) \begin{Bmatrix} \cos m'\theta \\ \sin m'\varphi \end{Bmatrix} r^2 \sin\theta \, d\theta \, d\varphi = \begin{cases} 0 & n \neq n' \text{ or } m \neq m' \\ \frac{1}{2n+1} & n = n' \text{ and } m = m \end{cases}$$

(7)

For example, the normalized surface harmonics for n=0, m=0 is 1,

for n=1, m=0 is $\cos\theta$, for n=1, m=1 is $\sin\theta \begin{Bmatrix} \cos \\ \sin \end{Bmatrix} \varphi$,, etc.

Different types of surface harmonics can be deduced from the

nature and forms of the normalized term: $P_n^m(\theta) \begin{Bmatrix} \cos m\varphi \\ \sin m\varphi \end{Bmatrix}$. If m = 0, the surface harmonic depends on colatitude, θ of latitude $(90° - \theta)$ as the longitude component vanishes. This surface harmonic is called the zonal harmonic. If n-m = 0, the surface harmonics depends on longitude and is called the sectoral harmonic (resembles the sectors of an orange). If m > 0 and n-m > 0, the harmonic is termed as tesseral harmonic. These harmonics are useful in characterizing the relative importance of coefficients A_n^m and B_n^m in equations (5) and (6).

If sources exist both inside and outside the sphere, then the potential, V in source-free regions near the surface of the sphere is given by the sum of the equations (5) and (6). Thus it is further convenient to express the combination of equations (5) and (6) in terms of Gauss' coefficients g_n^m and h_n^m for free-space (where the permeability, $\mu_0 = 1$ c.g.s.). Thus

$$V = a \sum_{n=1}^{\infty} \left(\frac{a}{r}\right)^{n+1} \sum_{m=0}^{n} \left[g_n^m \cos m\varphi + h_n^m \sin m\varphi \right] P_n^m(\theta)$$

(8)

It is generally known that n = 1 harmonic from equation (8) gives the first three coefficients $(g_1^0, g_1^1 \text{ and } h_1^1)$ which have overwhelming dominance. There is no g_0^0 term as this corresponds to the potential of a monopole which must therefore be zero. The first degree harmonic describes the potential of a dipole at the centre of the sphere and therefore the large amplitudes of these coefficients reflect the generally geocentric dipolar character of the main geomagnetic field.

Excluding the n = 1 harmonic from equation (8) eliminates the dipole term from the geomagnetic field and leaving a remainder of the form called the non-dipole part.

At the point of observation, P, T is the magnitude of the total field intensity, and X, Y, Z and H are the north, east, vertical and horizontal components respectively. The quantity, I is the angle T makes with the horizontal (along which H is directed) and is called the dip or inclination, while D, the declination is the angle the horizontal field, H makes with the true or geographic north. Note that H is not the same here as the one expressed in equation (2).

We note that a simple dipole theory predicts that the magnetic inclination, I is related to the geographic latitude, φ as $\tan I = 2\tan \varphi$.

The vector elements of the Earth's magnetic field at a point are

$\overline{T}, \overline{X}, \overline{Y}, \overline{Z}, D, I$. Like the reference ellipsoid and theoretical gravity, the mathematical representation of the low-degree parts of the geomagnetic field is determined by international agreement. This mathematical description is called the International Geomagnetic Reference Field (IGRF) and is attributed to the International Association of Geomagnetism and Aeronomy (IAGA) and its umbrella organization, the International Union of Geodesy and Geophysics (IUGG).

The IGRF is essentially a set of Gaussian coefficients, g_n^m and h_n^m that are put forth every 5 years by IAGA for use in a spherical harmonic model. At each of these epoch years, the group considers several proposals and typically adopts a compromise that best fits the data available. The coefficients for a given epoch year are referred to by IGRF and then the year, as in IGRF2000. The model includes both the coefficients for the epoch year and secular variation variables, which track the change of these coefficients in nanotesla per year. These secular variation coefficients are used to extrapolate the Gaussian coefficients to the date in question. Once data become available about the actual magnetic field for a given epoch year, the model is adjusted and becomes the Definitive Geomagnetic Reference Field, or DGRF.

Practically the IGRF consists of Gauss' coefficients through degree and order 10 or slightly above as these terms are believed to represent the larger part of the field of the Earth's core. Subtracting these low-order terms from the measured magnetic fields provides in principle the magnetic field of the crust.

SIMILARITIES AND DIFFERENCES WITH GRAVITY METHODS

The gravity and magnetic survey methods exploit the fact that variations in the physical properties of rocks in-situ give rise to variations in some physical quantity which may be measured remotely (on are above the ground). In the case of gravity method, the physical rock property is density and so density variations at all depths within the Earth contribute to the broad spectrum of gravity anomalies. For the magnetic method, the rock property is magnetic susceptibility and/or remanent magnetization; both of which can only exist at temperatures cooler than the Curie point and thus restricting the sources of magnetic anomalies to the uppermost 30 – 40 km of the Earth's interior. In practice, almost all magnetic properties of rocks in bulk reflect the properties and concentrations of oxides of iron and titanium (Fe and Ti): the Fe-Ti-O system, plus one sulphide mineral, pyrrhotite [1]. We also note that the highest density used typically in gravity surveys are about 3.0 g cm^{-3}, and the lowest densities are about 1.0 g cm^{-3}. Thus densities of rocks and soils vary very little from place to place. On the other hand, magnetic susceptibility can vary as much as four to five orders of magnitude from place to place, even within a given rock type.

Table 1: Other similarities and differences between gravity and magnetic methods

Magnetic method	Gravity method
Passive and is a potential field bearing all the consequences	Passive and is a potential field bearing all the consequences
Mathematical expression for the force field is that of the inverse square law relation	Mathematical expression for the force field is that of the inverse square law relation
Force between monopoles can either be attractive or repulsive	Force between masses is always attractive
A monopole cannot be isolated. Monopoles always exist in pairs (dipole)	A single point mass can be isolated

A properly reduced field has variation due to variation in induced magnetization of susceptible rocks and remanent magnetization	A properly reduced field has variation due to density variation in rocks
Field changes significantly over time (secular variation).	Field does not change significantly over time.

THE MAGNETIC PROPERTIES OF ROCKS

Geologic interpretation of magnetic data requires the knowledge of the magnetic properties of rocks in terms of magnetic susceptibility and remanent magnetization. Factors that influence rock magnetic properties for various rock types have been summarized appropriately [10], [1], [11], [12]. The rocks of the Earth's crust are in general only weakly magnetic but can exhibit both induced and remanent magnetizations. Magnetic properties of rocks can only exist at temperatures below the Curie point. The Curie temperature is found to vary within rocks but is often in the range 550°C to 600°C [13]. Modern research indicates that this temperature is probably reached by the normal geothermal gradient at depths between 30 and 40 km in the Earth and this so-called 'Curie point isotherm' may occur much closer to the Earth's surface in areas of high heat flow.

Indeed rock magnetism is a subject of considerable complexity. Clearly, all crustal rocks find themselves situated within the geomagnetic field described in section 3. These crustal rocks are therefore likely to display induced magnetization given by equation (3), where the magnitude of magnetization, J_i is proportional to the strength of the Earth's field, T. The magnetic susceptibility, k is actually the magnetic volume susceptibility that is encountered in exploration rather than mass or molar susceptibilities.

Apart from the induced magnetization, many rocks also show a natural remanent magnetization (NRM) that would remain even if the present-day geomagnetic field ceases to exist. The simplest way in which NRM can be acquired is through the process of cooling of rocks in molten state. As the rocks cool past the Curie point (or blocking temperature) a remanent magnetization in the direction of the prevailing geomagnetic field will be acquired. The magnitude and direction of the remanent magnetization can remain unchanged regardless of any subsequent changes in the ambient field.

MEASUREMENT PROCEDURES OF MAGNETIC FIELD

Measurements can be made of the Earth's total magnetic field or of components of the field in various directions. The oldest magnetic prospecting instrument is the magnetic compass, which measures the field direction. Other instruments include magnetic balances and fluxgate magnetometers.

The most used instruments in modern magnetic surveys are the proton-precession or optical-pumping magnetometers and these are appreciably more accurate and all of these instruments give absolute values of field. The proton magnetometer measures a radio-frequency voltage induced in a coil by the reorientation (precession) of magnetically polarized protons in a container of ordinary water or paraffin. Its measurement sensitivity is about 1 nT. The optical-pumping magnetometer makes use of the principles of nuclear resonance and cesium or rubidium vapour. It can detect minute magnetic fluctuations by measuring the effects of light-induced (optically pumped) transitions between atomic energy levels that are dependent on the strength of the prevailing magnetic field. The sensitivity of the optical absorption magnetometer is about 0.01 nT and on this premise may be preferred to proton precession magnetometer in air-borne surveys.

Airborne magnetic surveys or aeromagnetic surveys are usually made with magnetometers carried by aircraft flying in parallel lines spaced 2 - 4 km apart at an elevation of about 500 m when exploring for petroleum deposits and in lines 0.5 - 1.0 km apart roughly 200 m or less above the ground when searching for mineral concentrations. Ship-borne magnetic surveys or marine magnetic surveys can also be completed over water by towing a magnetometer behind a ship.

Ground surveys are conducted to follow up magnetic anomaly discoveries made from the air. Such surveys may involve stations spaced from 50 m apart. Survey may be along profiles or gridded network or may be in random pattern. Magnetometers also are towed by research vessels or mounted on the researcher on foot. In some cases, two or more magnetometers displaced a few metres from each other are used in a gradiometer arrangement; differences between their readings indicate the magnetic field gradient. A ground monitor is usually used to measure the natural fluctuations of the Earth's field over time so that corrections similar to drift correction in gravity can be made. Alternatively, like gravity observations where the temporal variation in field values were accounted for by reoccupying a base station and using the variation in this reading to account for instrument drift and temporal variations of the field, we could also use the same strategy in acquiring magnetic observations. The alternative is not the best as field variation in magnetic may be highly erratic and magnetometers which are electronic instruments do not drift. With these points in mind most investigators conduct magnetic surveys using two magnetometers. One is used to monitor temporal variations of the magnetic field continuously at a chosen "base station", and the other is used to collect observations related to the survey proper.

Surveying is generally suspended during periods of large magnetic fluctuation (magnetic storms).

MAGNETIC ANOMALIES

Magnetic effects result primarily from the magnetization induced in susceptible rocks by the Earth's magnetic field. Most sedimentary rocks have very low susceptibility and thus are nearly transparent to magnetism. Accordingly, in petroleum exploration magnetics are used negatively: magnetic anomalies indicate the absence of explorable sedimentary rocks. Magnetics are used for mapping features in igneous and metamorphic rocks, possibly faults, dikes, or other features that are associated with mineral concentrations. Data are usually displayed in the form of a contour map of the magnetic field, but interpretation is often made on profiles.

The first stage in any ground magnetic survey is to check the magnetometers and the operators. Operators can be potent sources of magnetic noise. Errors can also occur when the sensor of the magnetometer is carried on a short pole or on a back rack. Compasses, pocket knives, metal keys, geological hammers, and cultural articles with metal blend (belt, shoes, bungles, etc) are all detectable at distances below about a metre and therefore the use of high-sensitivity magnetometers requires that operators divest themselves of all metallic objects. Attempts must be made to follow the operation manual provided along with a magnetometer!

Diurnal corrections are essential in most field work, unless only gradient data are to be used. If only a single magnetometer is available, diurnal corrections have to rely on repeated visits to a base, ideally at intervals of less than an hour. A more robust diurnal curve can be constructed if a second fixed magnetometer is used to obtain readings at 3 to 5 minute intervals. The second magnetometer need not be the same type as that being used in the field. Thus a proton magnetometer can provide adequate diurnal control for surveys conducted with cesium vapour magnetometer and vice versa. Note that base should be remote from magnetic interferences and must be describable for future use.

In aeromagnetic surveys, great pains must be taken to eliminate spurious magnetic signals that may be expected to arise from the aircraft itself. Airframes of modern aircraft are primarily constructed from aluminum alloys which are non-magnetic and so the potential magnetic sources are the aircraft engines. Thus magnetometer sensors must be mounted far away from these engines.

Aeromagnetic data usually obtained have gone through on-board processing such as magnetic compensation, checking/editing, diurnal removal, tie line and micro leveling. For example, the basis of magnetic compensation is the reduction of motion-induced noise on the selected magnetic elements. These can be from individual sensors or various gradient configurations. The motion noise comes from the complex three-dimensional magnetic signature of the airframe as it changes attitude with respect to the magnetic field vector. The noise comes from permanent, induced and eddy effects of the airframe plus additional heading effects of the individual sensors. Thus the magnetic interference in a geophysical aircraft environment comes from several sources which must be noted and compensated for.

On-board data checking and editing involves the removal of spurious noise and spikes from the data. Such noise can be caused by cultural influences such as power lines, metallic structures, radio transmissions, fences and various other factors. Diurnal removal corrects for temporal variation of the earth's main field. This is achieved by subtracting the time-synchronized signal, recorded at a stationary base magnetometer, from survey data. Alternatively, points of intersection of tie lines with traverse/profile lines can form loop networks which can be used to correct for the diurnal variation similar to drift correction in gravity survey.

Tie leveling utilizes the additional data recorded on tie lines to further adjust the data by consideration of the observation that, after the above corrections are made, data recorded at intersections (crossover points) of traverse and tie lines should be equal. Several techniques exist for making these adjustments and [14] gives a detailed of the commonly used techniques. The most significant cause of these errors is usually inadequate diurnal removal. Micro-leveling, on the other hand, is used to remove any errors remaining after the above adjustments are applied. These are usually very subtle errors caused by variations in terrain clearance or elevated diurnal activity. Such errors manifest themselves in the data as anomalies elongate in the traverse line direction. Accordingly they can be successfully removed with directional spatial filtering techniques [15].

When all the above considerations to raw magnetic data have been recognized and attended to, the IGRF correction (main field effect) is now carried out to give the 'magnetic anomaly' defined as the departure of the observed field from the global model.

POTENTIAL FIELDS AND MODELS

Potential Field in Source Free Space

The potential field, φ (x, y, z) in free space, i.e. without any sources satisfies the Laplace equation $\nabla^2\varphi = 0$ or when sources are present the potential satisfies the so-called Poisson equation, $\nabla^2\varphi = -\rho(x, y, z)$, where ρ (x, y, z) stands for density or magnetization depending upon whether φ stands for gravity or magnetic potential. Laplace's equation has certain very useful properties particularly in source-free space such as the atmosphere where most measurements are made. Some of these properties are (1) given the potential field over any plane, we can compute the field at almost all points in the space by analytical continuation; (2) the points where the field cannot be computed are the so-called singular points. A closed surface enclosing all such singular points also encloses the sources which give rise to the potential field. These properties of the potential field are best brought out in the Fourier domain.

The Fourier transform in one-dimension can be found in most text books of applied mathematics. The Fourier transform in two or three dimensions possess additional properties worth noting [16]. The two-dimensional Fourier transform is given by:

$$\phi(u,v) = \iint_{-\infty}^{+\infty} \varphi(x,y)e^{-i(ux+vy)}$$

and its inverse is given by

$$\varphi(x,y) = \frac{1}{4\pi^2} \iint_{-\infty}^{+\infty} \phi(u,v)e^{i(ux+vy)}dudv .$$

Where u and v are spatial frequency numbers in the x- and y-directions respectively ($u = 2\pi/L_x$ and $v = 2\pi/L_y$), with L_x and L_y as length dimensions in the x- and y-directions respectively. It is important to note that φ (x, y) and Φ (u, v) are simply different ways of looking at the same phenomenon. The Fourier transform maps a function from one domain (space or time) into another domain (wave number or frequency). For details, [17] can be consulted.

Magnetic potential field is caused by the variation in magnetization in the Earth's crust. This potential field is observed over a plane close to the surface of the Earth. If the magnetization variations are properly modeled consistent with other geological information, it is possible to fit the model to the observed potential field. Note that the magnetic field induction usually observed is the derivative of this potential. The model parameters (body shape factors, susceptibility values, burial depth, and magnetization direction) are then observable. These models may be (1) excess magnetization confined to a well-defined geometrical object, (2) geological entity such as basins (sedimentary or metamorphic with intrusive bodies). Sedimentary basins are of great interest on account of their hydrocarbon potential and since these rocks are generally non-magnetic, the observed magnetic field is probably entirely due to the basement on which sediments are resting and (3) with available resources and technology (as in airborne magnetic surveys) large areas can be covered in a survey and so permitting maps that cover several geological provinces or basins and therefore allow inter basin studies such as delineating of extensive shallow and deep features as faults, basin boundaries, etc. to be extracted.

We shall briefly outline a few examples of the rigors that an interpreter goes through to synthesize information from these potential fields.

Dipole Field

We consider two monopoles of opposite sign: one at the origin of the 3-coordinate system and the other positioned below such that their common axis is along the z-axis, with $-\Delta z$ as the separation between the monopoles (Fig. 2)

The potential at P, V (P) due to both monopoles is the sum of the potentials caused by each monopole. This is given generally for monopoles that are not aligned along any particular axis as [18]:

$$V(P) = -C_m \vec{m}.\vec{\nabla}_P\left(\frac{1}{r}\right)$$

(9)

Where $\vec{m}$ is the dipole moment ($\vec{m} = \overline{qds}$ with q as the pole strength and $\overline{ds}$ is the displacement from monopole 1 to monopole 2.), C_m is the Coulomb's law constant ($C_m = \mu_0/4\pi$: μ_0 is the permeability of free space) and $\vec{\nabla}$ P is the derivative vector operator towards point P.

According to Helmholtz theorem, the magnetic field, $\vec{B}$ can be derived from this potential, V (P) such that $\vec{B} = -\vec{\nabla} P_V(P)$. Using this on equation (9) yields

$$\vec{B} = C_m \frac{m}{r^3}[3(\hat{m}.\hat{r})\hat{r} - \hat{m}], \quad r \neq 0$$

(10)

Where m is the magnitude of the dipole moment while $\hat{r}$ and $\hat{m}$ are unit vectors in the increasing r and m directions respectively. Thus equation (10) shows that the magnitude of $\vec{B}$ is proportional to the dipole moment and inversely proportional to the cube of the distance to the pole.

Equation (10) can also be expressed in cylindrical coordinates as [8]:

$$\vec{B} = C_m \frac{m}{r^3}(3\cos\theta \hat{r} - \hat{m})$$

(11)

Where θ is now the angle between $\hat{m}$ and $\hat{r}$ so that with $\vec{B} = -\vec{\nabla}_p V$, we can compute the components of the field in the directions of r and θ, respectively as $\vec{B}_r$ and $\vec{B}_\theta$ expressed as

$$\vec{B}_r = -\hat{r}\frac{\partial V}{\partial r} = 2C_m\frac{m\cos\theta}{r^3}, \vec{B}_\theta = \hat{\theta}\frac{1}{r}\frac{\partial V}{\partial r} = C_m\frac{m\sin\theta}{r^3}$$ and so the

magnitude of $\vec{B}$ is expressed as

$$|\vec{B}| = \sqrt{|\vec{B}_r|^2 + |\vec{B}_\theta|^2} = C_m\frac{m}{r^3}(3\cos^2\theta + 1)^{1/2}$$

(12)

Equation (12) shows that the magnitude $|\vec{B}|$ of the dipole field along any direction extending from the dipole decreases at a rate inversely proportional to the cube of the distance to the dipole. The magnitude of $\vec{B}$ also depends on θ.

Many magnetic bodies exist that are dipolar in nature to a first approximation. For example, the entire field of the Earth appears nearly dipolar from the perspective of other planets. It is also known that in aeromagnetic survey, the inhomogeneity of a massive pluton at the survey height appears to be a dipole source.

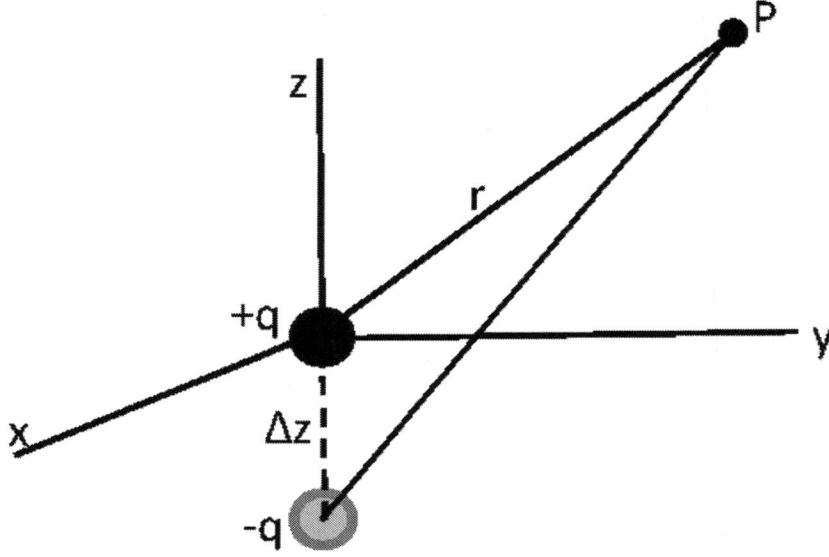

Figure 2: Two monopoles of opposite sign with the monopole of positive sign at the origin and other situated at $z = \Delta z$.

Three-Dimensional Distribution of Magnetization

We can consider a small element of magnetic material of volume, dv and magnetization, $\vec{M}$ (Fig. 3)to act like a single dipole such that $\vec{M}dv = \vec{m}$. Then the potential at some point, P outside the body is given as in equation (9) to be

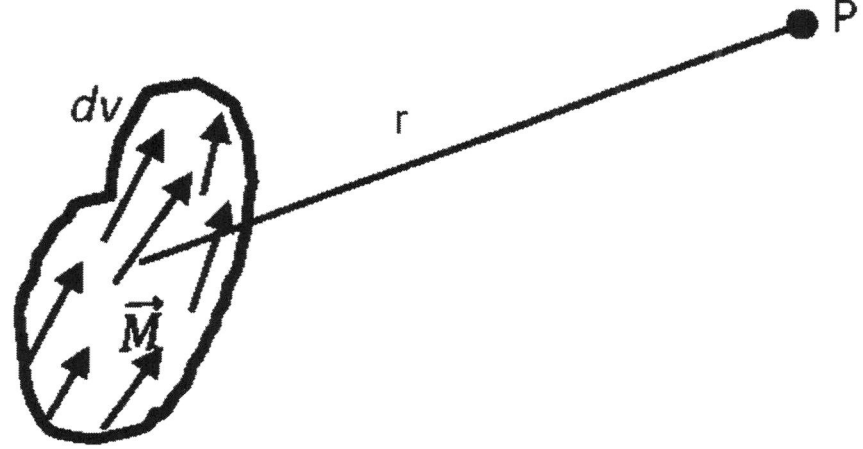

Figure 3: A 3-D magnetic body of volume dv and uniform magnetization $\vec{M}$.

$$V(P) = -C_m \vec{M} . \vec{\nabla}_P \left(\frac{1}{r}\right) dv$$

(13)

Where r again is the distance from P to the dipole. In general, magnetization, $\vec{M}$ is a function of position where both direction and magnitude can vary from point to point, i.e. $\vec{M} = \vec{M}$ (Q), where Q is the position of the volume element, dv. Integrating equation (13) over all elemental volumes provides the potential of a distribution of magnetization as

$$V(P) = C_m \int \vec{M}(Q) . \vec{\nabla}_Q \left(\frac{1}{r}\right) dv$$

(14)

The magnetic induction, $\vec{B}$ at P is then given by

$$\vec{B}(P) = -\vec{\nabla}_P V(P) = -C_m \vec{\nabla}_P \int \vec{M}(Q).\vec{\nabla}_Q \left(\frac{1}{r}\right) dv \tag{15}$$

The subscripts in the gradient operator from P to Q are now $\vec{\nabla}_P$ and $\vec{\nabla}_Q$ when the operator is inside the volume integral showing that the gradient is taken with respect to the source coordinates rather than with respect to the observation point.

For a two-dimensional source, we may start with a body of finite length 2a and so the volume integral in equation (15) is replaced with surface integral over the cross sectional area, dS of the body and a line integral along its length (the z-axis) as:

$$V(P) = C_m \int_r \vec{M}(Q).\vec{\nabla}_Q \left(\frac{1}{r}\right) dv = C_m \int_r \vec{M}(Q).\left(\int_{-a}^{a} \vec{\nabla}_Q \left(\frac{1}{r}\right) dz \right) dS \tag{a}$$

$$V(P) = 2C_m \int \frac{\vec{M}(Q).\hat{r}}{r} dS \tag{b}$$

$$\tag{16}$$

Where S is the cross sectional area of the body. As $a \to \infty$ (a 2-D case), the integral approaches a potential of infinite line of dipoles of unit magnitude. Hence (16b).

The magnetic field induction, $\vec{B}(P)$ can be obtained from equation (16b) as:

$$\vec{B}(P) = -\vec{\nabla} V(P) = 2C_m \int \frac{M(Q)}{r^2} [2(\hat{M}.\hat{r})\hat{r} - \hat{M}] dS \tag{17}$$

Note also that $\hat{r}$ is the normal to the long axis of the cylinder and r is a perpendicular distance.

Poisson Relation

We had noted in section one, and by implication, that the mutual force, F between a particle of mass m_1 centred at point Q (x', y', z') and a particle of mass, m_2 at P (x, y, z) is given by

$$F = G \frac{m_1 m_2}{r^2}$$

(18)

Where r= $[(x-x')^2+(y-y')^2+(z-z')^2]^{1/2}$. If we let m_2 be a test particle with unit magnitude, then the gravitational attraction, $\vec{g}$ (P) produced by m_1 at the location P of m_2 (the test particle) is

$$\vec{g}(P) = -G \frac{m_1}{r} \hat{r}$$

(19)

Where $\hat{r}$ is a unit vector directed from m_1 to the observation point, P and expressed as $\hat{r} = \frac{1}{r}\left[(x - x')\hat{i} + (y + y')\hat{j} + (z - z')\hat{k}\right]$ and the minus sign in equation (19) indicates that as r increases $\hat{g}$ (P) decreases in absolute value.

If a potential exists, then the gravitational attraction (also known as the gravitational acceleration), $\hat{g}$ (P) can be derived from this potential. Let this potential be U. Then at P, U = U (P), such that

$$\vec{g}(P) = -\vec{\nabla} U(P)$$

(20)

Where here, U (P) can be expressed as

$$U(P) = G \frac{m_1}{r^2}$$

(21)

Where the gravitational potential is defined as the work done by the field on the test particle.

Equations (19) and (12) show that the magnetic scalar potential of an element of magnetic material and the gravitational attraction of mass are identical. Starting from equation (13), and considering a body with uniform magnetization, $\overline{M}$ and uniform density, ρ, the magnetic scalar potential at a point, P is

$$V(P) = -C_m \int \overrightarrow{M} . \overrightarrow{\nabla}_P \left(\frac{1}{r}\right) dv = -C_m \overrightarrow{M} . \overrightarrow{\nabla}_P \int \frac{1}{r} dv \tag{22}$$

The gravitational potential (equation (22)) is written as

$U(P) = G \int \frac{\rho}{r} dv = G\rho \int \frac{1}{r} dv$, so that $\int \frac{1}{r} dv = \frac{U(P)}{G_\rho}$ and substituting this in equation (21) gives

$$V(P) = -\frac{C_m}{G\rho} \overrightarrow{M} . \overrightarrow{\nabla}_P U(P) = -\frac{C_m M}{G\rho} g_m \tag{23}$$

Where gm is the component of gravity in the direction of magnetization and M is the magnitude of the magnetization.

Equation (23) is the so-called Poisson's relation and can be stated as follows: if the boundaries of a gravitational and magnetic source are the same and its magnetization and density distribution being uniform, then the magnetic potential is proportional to the component of gravitational attraction in the direction of magnetization, i.e. V (P)∝ $\overline{M} . \overrightarrow{g}$ (P).

Poisson relation can be used to transform a magnetic anomaly into pseudo-gravity, the gravity anomaly that would be observed if the magnetization were replaced by a density distribution of exact proportions [19]. Pseudo-gravity transformation is a good aid in interpretation of magnetic data. In addition, Poisson's relation can be used to derive expressions for the magnetic induction of simple bodies when the expression for gravitation attraction is known.

Magnetic Field over Simple Geometrical Bodies

The form of magnetic anomaly from a given body is complex and generally depends on the following factors:

- The geometry of the body
- The direction of the Earth's field at a location of the body
- The direction of polarization of the rocks forming the body
- The orientation of the body with respect to the direction of the Earth's field
- The orientation of the line of observation with respect to the axis of the body.

Thus the computations of models to account for magnetic anomalies are much more complex than those for gravity anomalies. As earlier stated, when the gravity expressions for simple geometrical bodies are given, we can use the Poisson's relation to find the magnetic expressions over these models [18]. Table 2 is a summary of few of such computations.

Table 2: Gravity and magnetic potentials caused by simple sources, along with magnetic induction for bodies of uniform density, ρ and magnetization, $\overrightarrow{M}$ observed at some point, P (x, y, z) away from the centre of the body (other symbols are defined as in section 8.2)

Shape of Body	Gravity potential, U	Magnetic potential, V	Magnetic field, $\overrightarrow{B} = \nabla V$
Sphere of radius, a	$\frac{4}{3}\pi a^3 \frac{G\rho}{r}$	$C_m \frac{\overrightarrow{m}.\hat{r}}{r^2}$	$C_m \frac{m}{r^3}[3(\hat{m}.\hat{r})\hat{r} - \hat{m}]$

Horizontal cylinder of infinite length of cross sectional radius, r	$2\pi a^2 \rho G \log\left(\frac{1}{r}\right)$	$2C_m\pi a^2\dfrac{\vec{M}.\hat{r}}{r}$	$\frac{2C_m m'}{r^2}[2(\hat{m}'.\hat{r})\hat{r} - \hat{m}']$
Horizontal slab of thickness, t	$2\pi\rho Gtz$	$-2\pi C_m Mt$	Zero

MAGNETIC DATA PROCESSING AND INTERPRETATION

The total-field magnetic anomaly of section 7 which was obtained by subtracting the magnitude of a suitable regional field (the IGRF or DGRF model for the survey date) from the total-field measurement is referred to as the crustal field. As earlier stated, this field is a vector sum of the remanent and the induced fields of the magnetically susceptible rocks of the crust down to the bottom of the Curie depth. The induced field component is usually in the same direction as the ambient field during the survey period.

Magnetic data processing includes everything done to the data between acquisition and the creation of an interpretable profile map or digital data set. Apart from the effect earlier discussed which are ignored or avoided, rather than corrected for, the correction required for ground magnetic data are insignificant especially when compared to gravity. The influence of topography (terrain) on ground magnetics on the other hand can be significant. Magnetic terrain effects can severely mask the signatures of underlying sources as demonstrated by [20]. Many workers have attempted to remove or minimize magnetic terrain effects by using some form of filtering as summarized in [21].

Interpretation of magnetic anomalies has to do with (a) studying the given magnetic map, profile or digital data to have a picture of the probable subsurface causes (qualitative interpretation), (b) separating the effect of individual features on the basis of available geophysical and geological data (further separation of broad-based or long-wavelength anomalies) and (c) estimating the likely parameters of the bodies of interest from the corresponding 'residual' or anomalies (quantitative interpretation).

The last two categories of interpretation procedures can be further broken into three parts. Each part has the goal of illuminating the spatial distribution of magnetic sources, but they approach the goal with quite different logical processes.

- *Forward Method*: an initial model for the source body is constructed based on geologic and geophysical intuition. The initial model anomaly is calculated and compared with the observed anomaly; and model parameters are adjusted in order to improve the fit between the two anomalies until the calculated and observed anomalies are deemed sufficiently alike. Forward method is source modeling in which magnetic anomalies were interpreted using characteristics curves [22] calculated from simple models (before the use of electronic computers) or using computer algorithms. Such schemes include 2-D magnetic models [23] and many workers that followed), 3-D magnetic models [24] and many other improvements that followed), Fourier-based models [25] and other improvements that followed) and voxel-based models [26] and others that followed).

- *Inverse Method*: one or more body parameters are calculated automatically and directly from the observed anomaly through some plausible assumptions of the form of the source body. Under this category, we have depth-to-source estimation techniques such as Werner deconvolution [27] and other workers that followed), frequency-domain techniques [28] and others that followed), Naudy matched filter based method [29] and others that followed), analytic signal method [30], [31] and others that followed), Euler deconvolution [32] and others that followed), source parameter imaging [33] and others that followed) and statistical methods [34] and others that followed). Physical property mapping under the inverse method include terracing [35] and susceptibility mapping [36] and others that followed). Other inversion techniques have to do with automated numerical procedures which construct models of subsurface geology from magnetic data and any prior information [37], [38], [39], [40], [41] among many others).

- **_Data Enhancement and Display_**: model parameters are not calculated, but the anomaly is processed in some way in order to enhance certain characteristics of the source bodies, thereby facilitating the overall interpretation. This category involves all filter-based analyses such as reduction to pole (RTP) and pseudogravity [19] and others that followed), upward/downward continuation [42] and others that followed), derivative filters [43] and many others), matched filtering [44] and others that followed) and wavelet transforms [45] and others that followed). Some of the enhancement techniques are artificial illumination [46], automatic gain control [47] and textural filtering [48]. Data displays can be in form of stacked profiles, contour maps, images and bipole plots.

In summary, any geophysical survey has two domains or spaces of interest: data space (data collected from the field) and model space (earth models) that account for the data space. The task is to establish a link between a data space and a model space (Fig.4).

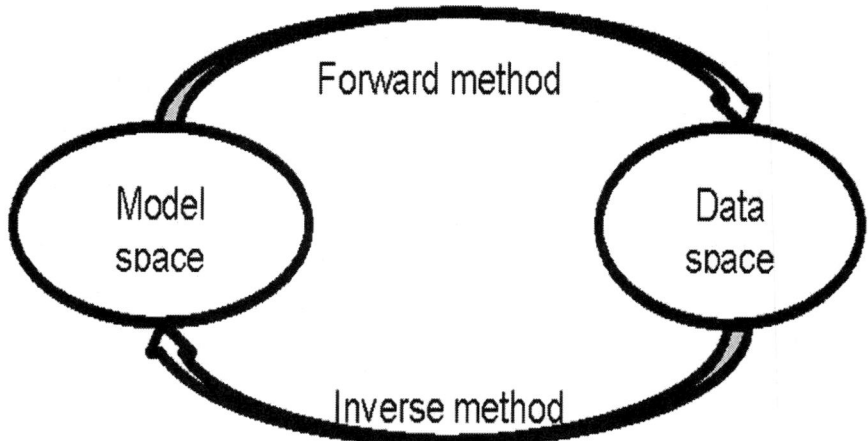

Figure 4: Connecting link between data space and model space in forward and inverse method.

The task of retrieving complete information about model parameters from a complete and precise set of data is inversion. Thus if we have a set of data collected from the field, we try to say about the earth model with those finite data set. How many different ways can one try to travel within a data space and a model space? What difficulties are encountered? How many different ways can one try to overcome those difficulties? How much information can one really gather and what are their limitations? What precautions should be taken as one moves from one space to another? Inverse theory seems to address these questions. Inverse method is a direct method in which source parameters are determined in a direct way from field (e.g. magnetic field) measurements.

Forward method on the other hand entails starting from model space (Fig. 4) by guessing initial model parameters and then calculating the model anomaly (in data space). The model anomaly is compared with observed (data) anomaly. If the match between the two is acceptable, the process stops, otherwise model parameters are adjusted and the process repeated. Forward method is an indirect method.

Various formulae exist for computing magnetic field of regular shapes such as the ones given in Table 2.

Analysis of magnetic data and their various enhancements via a suite of qualitative and quantitative methods as outlined above results in an interpretation of the subsurface geology. Qualitative interpretation relies on the spatial patterns that an interpreter can recognize in the data. Faults, dykes, lineaments and folds are usually identified. Intrusive bodies are often recognized by virtue of the shape and amplitude of their anomalies and so on. For example, detection of a fault in a magnetic map is an important exercise in mineral prospecting. Usually faults and related fractures serve as major channel-ways for the upward migration of ore-bearing fluids. Fault zones containing altered magnetic minerals can be detected from series of closed lows on contour maps, by inflections or by lows on magnetic profiles or by displaced magnetic marker horizons.

Quantitative interpretation on the other hand is meant to compliment the qualitative method and seeks to provide useful estimates of the geometry, depth and magnetization of the magnetic sources. Broadly categorized as curve matching, forward modeling or inversion, quantitative techniques rely on the notion that simple geometric bodies, whose magnetic anomaly can be theoretically calculated, can adequately approximate magnetically more complex bodies. Geometric bodies such as ellipsoids, plates, rectangular prisms, polygonal prisms and thin sheets can all be calculated. Complex bodies can be built by superposing the effects of several simple bodies. Faults are often modeled using thin sheet model.

Like most other geophysical methods, magnetics is ambiguous to the extent that there are an infinite number of models that have the same magnetic anomaly. Acceptable models should be tested for geological plausibility.

CASE STUDY: AEROMAGNETIC FIELD OVER THE UPPER BENUE TROUGH, NIGERIA

Geological Framework

The Upper Benue Trough, Nigeria (UBT) is the northern end of the nearly 1000 – km NE-SW trending sediment-filled Benue Trough, Nigeria (Fig. 5). Apart from the adjacent Niger Delta area and the offshore region where oil/gas exploration and production are taking place, the Benue Trough as an inland basin lacks the same full attention of the oil/gas companies. However, with the upbeat in the exploration efforts of the national government towards the search for hydrocarbon prospects of this inland basin, particularly in the light of new oil discoveries in the nearby genetically related basins, attention is directed at the structural setting of this basin.

The Benue Trough as a NE-SW trending sedimentary basin has a Y – shaped northern end (a near E – W trending branch of the Yola – Garoua and north-trending branch of Gongola – Muri) (Fig. 5). The Benue Trough is filled with sediments that range from Late Aptian to Palaeocene in age and whose thickness could reach up to 6000 m at places. The environments of deposition also varied over time such that the sediments vary from continental lacustrine/fluviatile sediments at the bottom through various marine transgressive and regressive beds, to immature reddish continental sands at the top.

For the past 50 years, the published works on the geology of the Benue Trough have substantially increased. The most important regional geological work on the Benue Trough by [49] was a basis for subsequent geological interpretations. [49] interpreted the Benue Trough origin in terms of rift faulting and the folding of the Cretaceous age associated with the basement flexuring. The first geophysical contribution of [50] on the Benue Trough remains to date the unique published reference. These authors have proposed the same rift origin considering that the main boundary rift faults are concealed by the Cretaceous sediments. They observed the existence below the Benue Trough axis, of a central positive gravity anomaly interpreted as a basement high. Field evidence indicates that a set of deep-seated faults is superimposed on the axial high and controlled the tectonic evolution of the trough.

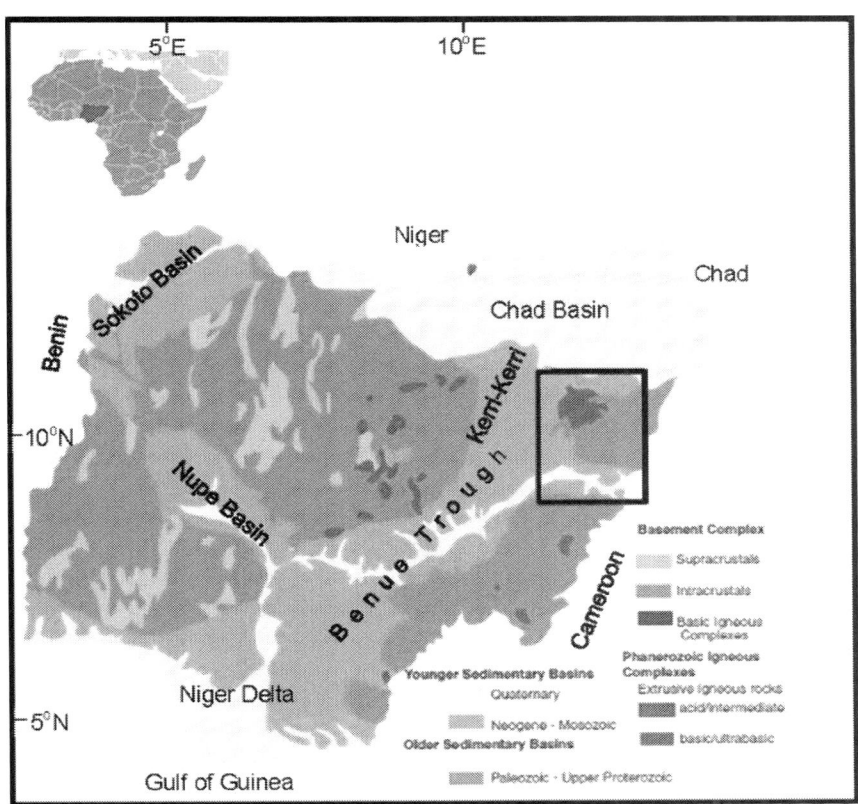

Figure 5: Top is map of Africa showing Nigeria. Below is a simplified geological map of Nigeria (modified from http://nigeriaminers, org). The inset rectangle is the Upper Benue Trough, Nigeria.

The rift origin of the Benue Trough supported by numerous authors was interpreted in the plate tectonics concept and from the 1970s several models were proposed to explain the origin of the Benue Trough. For example, seen as a direct consequence of the Atlantic Ocean opening, the Benue Trough was considered to be the third arm of a triple junction located beneath the centre of the present Niger Delta [51] and proposed a Ridge-Ridge-Ridge (RRR) triple junction. This hypothesis has been widely discussed and replaced in the general framework of African Rift System.

The Benue Trough is subdivided into 3 units on the basis of stratigraphic and tectonic considerations. The southern ensemble is called the Lower Benue Trough (LBT) and includes two main units: the Abakaliki Uplift or Anticlinorium and the Anambra Basin. The Abakaliki Anticlinorium (AA) is formed by tightly folded Cretaceous sediments intruded by numerous magmatic rocks. From the Niger Delta, AA extends for about 250 km to the Gboko – Ogoja area in a N50E direction. To the north of AA is a vast synclinorial structure called the Anambra Basin and trends in a N30E direction. This basin comprises a thick undeformed Cretaceous series. On the northern margin of the Anambra Basin, is the Nupe or Middle Niger Basin which stretches along a NW-SE direction. To the south, AA is flanked by the Afikpo Syncline and by a narrow strip of thin, undeformed sediments resting on the Basement Complex (the Mamfe Basin) and to the northwestern border is the Oban Massif. South of the Oban Massif is the Calabar Flank, which belongs to the coastal basins of the Gulf of Guinea.

The Upper Benue Trough (UBT) which is the northern ensemble is the most complex part (Fig. 5). It is characterized by cover tectonics and can be further subdivided into several smaller units. The Gongola – Muri and Yola – Garoua branches are digitations of the Benue Trough and present a similar tectonic style. The Gongola - Muri rift disappears beneath the Tertiary sediments of the Chad Basin and so the margins of this rift are geologically the most difficult to established. The Yola – Garoua rift to some extent is the least known of the West African Rift and strikes E-W into Cameroon. On the western margin of the UBT is the flat-lying Paleocene Kerri Kerri basin resting unconformably upon the folded Cretaceous. The development of the Kerri Kerri basin is said to be controlled by a set of faults between it and the Basement Complex of the Jos Plateau [52]. The basin formation and its tectonic activity seem to be a response of the general uplift of the UBT due to late Cretaceous folding.

The UBT is contiguous with the Nigerian sector of the Chad Basin which extends northwards into the Termit Basin of Chad and Niger and southwestwards into Cameroon and southern Chad as Bongor, Doba, Doseo and Salamat basins. This rift system is closely linked with oil-rich Muglad Basin of Sudan.

Aeromagnetic Field

Magnetic data over Nigeria have been largely collected above the ground surface in form of systematic surveys on behalf of the national government. These airborne surveys were carried out principally by a consultant, namely: **Fugro Airborne Surveys,** on behalf of the Nigerian Geological Survey Agency (NGSA) between 2003 and 2009. The main aim of these surveys has been to assist in mineral and groundwater development through improved geological mapping. Flight line direction is nearly NW-SE and tie line direction is NE – SW. The flight height average is 100 m; profile line spacing is 500 m with tie line spacing of 2 km. Figure 6 shows one of the aircrafts of Fugro Airborne Surveys in flight.

The total field aeromagnetic field intensity for the UBT comprises 16 half-degree grids acquired from NGSA and is used for the purpose of the present study. These are 131_BAJOGA, 132_GULANI, 133_BIU, 134_CHIBUK; 152_GOMBE, 153_WUYO, 154_SHANI, 155_GARKIDA; 173_KALTUNGO, 174_GUYOK, 175_SHELLEN, 176_ZUMO; 194_LAU, 195_DONG, 196_NUMAN and 197_GIREI total magnetic intensity (TMI) grids.

Figure 6: Fugro Airborne Surveys photo showing a magnetometer in a 'stinger' behind the fixed-wing aircraft.

The aeromagnetic data obtained have gone through on-board processing such as magnetic compensation, checking/editing, diurnal removal, tie line and micro leveling.

The composite total field aeromagnetic data for the UBT are displayed in image (Fig. 7). The advantage of images is that they are capable of showing extremely subtle features not apparent in other forms of presentations (such as contour maps). They can also be quickly manipulated in digital form, thereby providing an ideal basis for on-screen GIS-based applications.

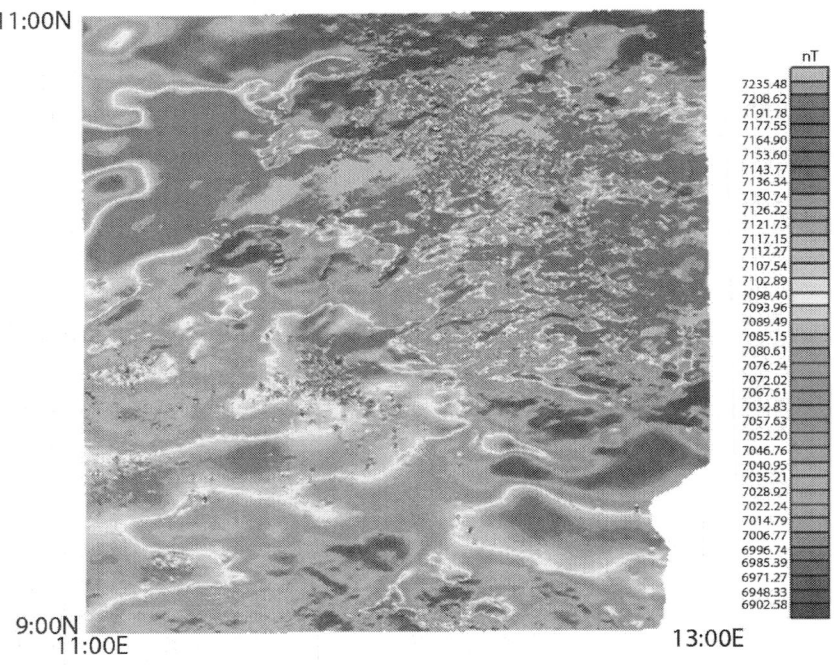

Figure 7: The total-field aeromagnetic intensity over UBT. A base value of 26, 000 nT should be added to map value for the total-field.

We have further treated the composite total field aeromagnetic data (Fig. 7) for the UBT for the main field effect through the removal of the Definitive Geomagnetic Reference Field (DGRF – 2005) (Fig. 8) resulting in total magnetic field intensity anomaly (Fig. 9). This anomaly field has polarity signs that shows the impact of low geomagnetic inclination values for the study area and also reflects (1) theinduced field caused by magnetic induction in magnetically susceptible earth materials polarized by the main field and (2) the field caused by remanent magnetism of earth materials. We call these two fields, the crustal field and have used the appropriate software (Geosoft Oasis Montaj version 8.0) for image processing and/or display of both the raw data (Fig. 7) and the anomaly data (Fig. 9). We have noted that a NW-SE mega feature dominated the middle of the study area. This linear feature, which is interrupted somehow towards the NW section of the map area, is believed to be central in the structural configuration and set-up of this Benue Trough subarea.

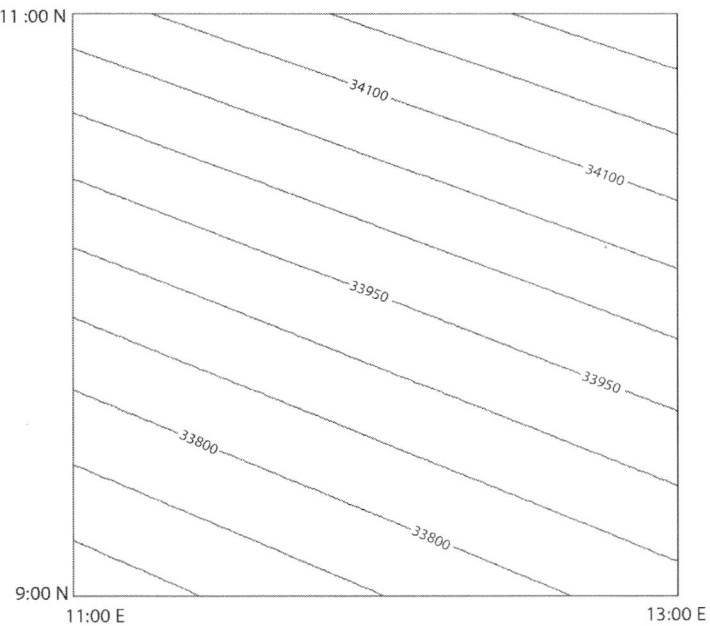

Figure 8: Definitive Geomagnetic Reference Model (DGRF2005) over UBT at 100 m above ground level. The Earth's model used is geodetic and CI is 50 nT.

Analysis of Aeromagnetic Field

We have computed for the present study area (UBT) the field, F for the epoch year 2005 and is displayed in contour form (Fig. 8) and because of the relatively small size of the study area, the values of D and I cannot be contoured and imaged. However between these values of Longitudes and Latitudes (11°E – 13°E, 9°N – 11°N), D ranges from -1.4° (11°E, 9°N) to -0.7° (13°E, 11°N) and I ranges from -5.7° (11°E, 9°N) to -0.2° (13°E, 11°N).

The Definitive Geomagnetic Reference Field (DGRF) or the main field map of the study area (Fig. 8)shows a NW-SE trending lines that have increasing values from the SW portion (minimum value of 33688 nT) to the NE portion (maximum value of 34271 nT) having an average value of 33974 nT and standard deviation of 129 nT. The inclination of the field for this epoch period (2005) decreases correspondingly from -5.7° to -0.2° indicating that slightly north of 11° latitude, the inclination of 0° or magnetic equator passes. We are therefore dealing with a low magnetic latitude area. Similarly the geomagnetic declination varies correspondingly from -1.4° to -0.7° which also shows that further north of 11° latitude, the declination would be 0°, indicating that geomagnetic and geographic meridians coincide. Computations of the rate of change of declination, D (in minutes per year) shows a constant value of 6 minutes/year, rate of change of inclination, I shows a northward increase from -4 to -3 minutes/year and also a northward increase of secular variation in the total field, F of between 21 – 24 nT/ year. This shows that beginning 2010, Nigeria will be completely in the southern magnetic hemisphere in the next 40 years, where then the 0° latitude or magnetic equator will be passing through Niger Republic.

The image display of the aeromagnetic total field anomaly map (Fig. 9) has negative anomaly values. This is not surprising and in fact it is expected. The study area and generally Nigeria is situated in a magnetically low-latitude area. The polarizing field of the Earth in such areas is the horizontal component, H. Note that the structure of the Earth' magnetic field resembles that of a bar magnet. At the magnetic poles, the field is essentially vertical (Z), at the centre of the magnetic bar the field is horizontal (H). The horizontal component, H of the total field, F around or at the magnetic equator is therefore the polarizing field. Any magnetically susceptible (non-zero susceptibility) earth materials within this area will be magnetized or polarized by H. When H field induces a polarization field in a susceptible material, the orientation of the field lines describing the magnetic field is rotated 90°. Above this susceptible earth material, the polarization field now points in the opposite direction as the Earth's main field. Thus the total field measured will be less than the Earth's main field, and so upon removal of the main field, the resulting anomalous field will be negative. This is not the case in high-latitude areas, for the same susceptible earth material, where the anomalous field over such would be largely positive and/or negative where also the rotation of the polarizing field depends on the value of inclination, I.

The anomaly map in Figure 9 can be broadly characterized into at least 4 colour zones with the following grid values: -2264 to – 982 nT, -982 to – 877 nT, -877 to -731 nT and -731 to -653 nT running from the NE edge and terminating to the SW side. There appears to be a shear zone running NW-SE nearly bisecting the area and passing through Girei, Shellen, Wuyo and disappearing or being interrupted by Gombe grid probably by reason of an offset NE-SW feature occupying the middle of Gombe grid and pinching out on Biu grid (Fig. 9). The Biu basalts and other high-susceptibility rocks around must have been very influential in the recorded low magnetic anomaly values at NE portion of the map area, particularly towards the northern edges of Bajoga, Gulani, Biu and Chibuk grids.

To demonstrate the usefulness of digital tools in the analysis of magnetic data, we shall apply only one digital processing tool to the analysis of the aeromagnetic total-field anomaly over UBT. We shall use the analytic signal technique.

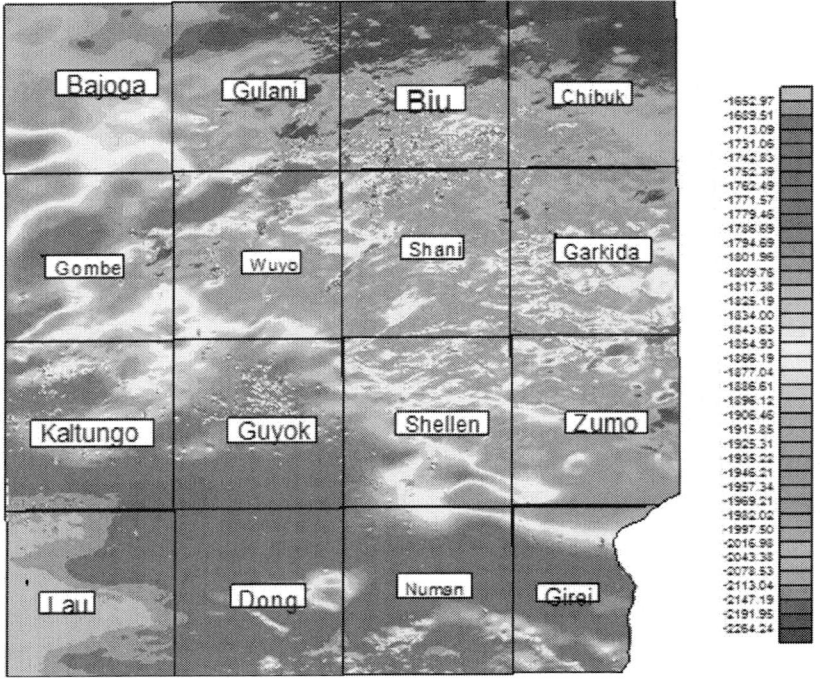

Figure 9: The total-field aeromagnetic anomaly over UBT. A value of 1000 nT should be added to map values.

The analytic signal for magnetic anomalies was initially defined as a 'complex field deriving from a complex potential' [30]. A 3-D analytic signal $\vec{A}$ [43], [53] of a potential field anomaly, M (magnetic field or vertical gradient of gravity), can be defined as:

$$\vec{A}(x, \; y) = \frac{\partial M}{\partial x}\hat{x} + \frac{\partial M}{\partial y}\hat{y} + \frac{\partial M}{\partial z}\hat{z}$$

$$(24)$$

Where $\hat{x}, \hat{y}, \hat{z}$ are unit vectors in the x-, y- and z-axis directions? The analytic signal amplitude or its absolute value can be expressed by a vector addition of the two real components in the x and y directions and of the imaginary component in the z-direction, i.e.

$$|\vec{A}(x,\ y)| = \sqrt{\left(\frac{\partial M}{\partial x}\right)^2 + \left(\frac{\partial M}{\partial y}\right)^2 + \left(\frac{\partial M}{\partial z}\right)^2}$$

(25)

The field and the analytic signal derivatives are more easily derived in the wave number domain. If F (M) is the Fourier transform of M in the 2-D wave number domain with wave number $\vec{k} = (k_x, k_y)$, the horizontal and vertical derivatives of M correspond respectively to multiplication of F (M) by $i(k_x, k_y) = i\vec{k}$ and $|\vec{k}|$ In 3-D the gradient operator in frequency domain is given by $\vec{\nabla} = ik_x\hat{x} + ik_y\hat{y} + |\vec{k}|\hat{z}$ The Fourier transform of the analytic signal can then be expressed in terms of the gradient of the Fourier transform of the field M by the following equation equivalent to the space-domain equation above, i.e. [53]:

$$\hat{t}.F\left(\vec{A}(x,y)\right) = \hat{h}.\vec{\nabla}F(M) + i\hat{z}\vec{\nabla}F(M)$$

(26)

Where $\hat{h} = \hat{x} + \hat{y}$ is the horizontal unit vector and $\hat{t} = \hat{h} + \hat{z}$.

The amplitude of the 3-D analytic signal of the total magnetic field anomaly produces maxima over magnetic contacts regardless of the direction of magnetization. The absence of magnetization direction in the shape of analytic signal anomalies is a particularly attractive characteristic for the interpretation of magnetic field near the magnetic latitude like the area under test. It is also known that the depths to sources can be determined from the distance between inflection points of analytic signal anomalies, but have not explored that option and interested readers can refer to [54].

In this method, we have applied the concept of analytic signal to the residual total magnetic field intensity of the UBT. These processes were accomplished by use of Geosoft Oasis Montaj (version 8.0).

Figure 10 shows the output of the analytical signal amplitude calculated from the original total-field magnetic anomaly (Fig. 9).

Analytic signal of the total-field magnetic anomaly reduces magnetic data to anomalies whose maxima mark the edges of magnetized bodies and whose shape can be used to determine the depth of these edges (we have not done this second aspect).

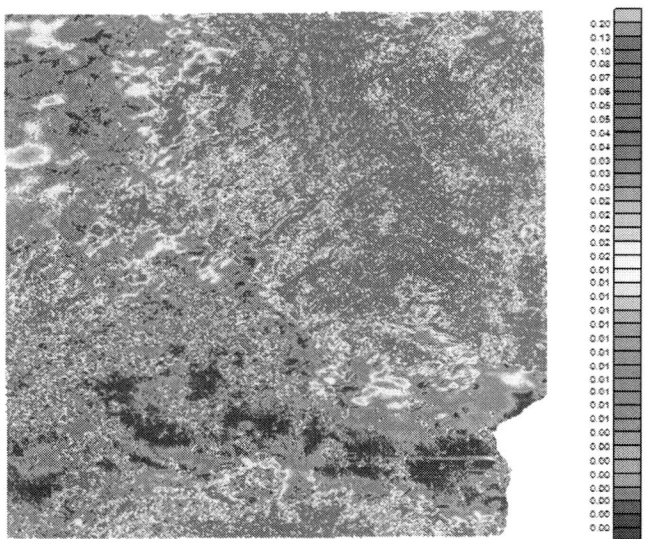

Analytic signal amplitude (nT/m)

Figure 10: Output of the analytic signal amplitude over UBT. The boundary of high amplitude anomaly over the Biu area (basaltic areas) are delineated.

The analytic signal amplitude over the UBT ranges from 0.00 nT/m to 7.93 nT/m: having a mean of 0.036 nT/m and standard deviation of 0.073 nT/m. Since amplitude of the analytic signal anomalies combines all vector components of the field into a simple constant, a good way to think of analytic signal is as a map of magnetization in the ground. With this in mind, we can picture strong anomalies to exist over where the magnetization vector intersects magnetic contrasts, even though one cannot know the source of the contrasts from the signal amplitude alone. Consequently, we can easily see the boundaries of the Biu basalts properly demarcated (Fig. 10) shown by the higher analytic signal values. Note also the few scattered imprints of the same basalts tailing to the SW direction from this major anomaly. The Biu area is composed of Tertiary and Quaternary periods (less than 65 Ma ago) basaltic lava flows containing abundant peridotite xenoliths.

CONCLUSIONS

In this chapter we have explored the magnetic method for economic exploration of the Earth. The strength of the method lies in the adequate distribution of magnetization within the crustal materials of the Earth in the light of measurable magnetic field over them.

The Earth's magnetic field, that is central in the remanence and induced processes, is itself complex. Spherical harmonic analysis provides the means of characterizing the Earth's magnetic field and with such a representation; it is possible to predict the geomagnetic dipolar field and other non-dipolar components. The knowledge of the dipolar field of the Earth enables the magnetic anomaly to be determined over a survey area from measurements of the magnetic field induction.

We have applied the magnetic method to real field measurements of total-field aeromagnetic intensity data over the Upper Benue Trough, Nigeria. The working data were corrected for secular variation using the existing DGRF model. The anomaly field which is the summary of the crustal field was further processed to obtain the amplitude of analytic signal of the anomaly field. The analytic signal transformations combine derivative calculations to produce an attribute that is independent of the main inclination and direction of magnetization as well as having peaks over the edges of wide bodies. Thus a simple relationship between the geometry of the causative bodies and the transformed data is observed such as seen in Figure 10. We note that the borders of the Biu basalt as exposed by the applied analytic signal technique of the magnetic anomaly data goes beyond the outcropping boundary that may be offered by remotely sensed data. We recognize that even though the magnetic data were remotely sensed, the result from such measurements goes beyond what the traditional remote sensing information can offer.

The greatest limitation of the magnetic method is the fact that it only responds to variations in the magnetic properties of the earth materials and so many other characteristics of the subsurface (e.g. regolith characteristics) are not resolvable. The inherent ambiguity in magnetic interpretation (for quantitative techniques) is problematic where several geologically plausible models can be attained from the data. Interpreters of magnetic data must therefore be aware of such limitations and be prepared to obtain confirmatory facts from other databases to decrease the level of ambiguity.

ACKNOWLEDGEMENTS

I acknowledge the National Centre for Petroleum Research and Development (NCPRD), of the Energy Commission of Nigeria (ECN), Abubakar Tafawa Balewa Balewa University, Bauchi, Nigeria for supporting this research.

REFERENCES

1.   Grant, F. S. Aeromagnetics, geology and environments I, magnetite in igneous, sedimentary and metamorphic rocks: an overview. Geoexploration1985; 23 303 - 333

2.   Campbell, W. C. Introduction to geomagnetic fields: Cambridge University Press; 1997

3.   Cox, A. Plate tectonics and geomagnetic reversals. W. H. Freeman and Company; 1973

4.   McElhinny, W. M. Paleomagnetics and plate tectonics: Cambridge University Press; 1973

5.   McElhinny, W. M., McFadden, P. L. Paleomagnetism – continents and oceans: Academic Press; 2000

6.   Merrill, R. T., McElhinny, M. W., McFadden, P. L. Magnetic field of the earth; paleomagnetism, the core and the deep mantle: Academic Press, San Diego;1996

7.   Doell, R., Cox, A. Magnetization of rocks, In: Mining Geophysics Volume 2: Theory, Society of Exploration Geophysicists SEG Mining Geophysics Volume Editorial Committee (Eds.):1967 446-453, Tulsa

8.   Telford, W. M., Geldart, L. P., Sheriff, R. E., Keys, D. A. Applied geophysics. Cambridge University Press, London:1990

9.   Chapman, S., Bartels, J. Geomagnetism. Oxford University Press:1940

10.   Haggerty, S. E. The aeromagnetic mineralogy of igneous rocks. Canadian Journal of Earth Sciences 1979: 16 1281-1293

11.   Reynolds, R. L., Rosenbaum, J. G., Hudson, M. R., Fishman, N. S. Rock magnetism, the distribution of magnetic minerals in Earth's crust and aeromagnetic anomalies. U. S. Geological Survey Bulletin 1990; 1924.24-45.

12. Clark, D. A. Magnetic petrophysics and magnetic petrology: aids to geological interpretation of magnetic surveys. AGSO Journal of Australian Geology and Geophysics 1997;17.83-103

13. Clark, D. A., Emerson, D. W. Notes on rock magnetization characteristics in applied geophysical studies. Exploration Geophysics 1991; 22.547-555.

14. Luyendyk, A. P. J. Processing of airborne magnetic data. AGSO J. Australian Geol. Geophys. 1998;17 31-38

15. Minty, B. R. S. Simple micro-leveling for aeromagnetic data. Exploration Geophysics 1991;.22.591-592.

16. Kay, S. M. Fundamentals of statistical signal processing. Prentice Hall, Englewood Cliffs, NJ.; 1993

17. Naidu, P. S., Mathew, M. P. Analysis of geophysical potential fields. Elsevier Science Publishers, Netherlands; 1998

18. Blakely, R. J. Potential theory in gravity and magnetic applications. Cambridge University Press; 1996

19. Baranov, V. A new method for interpretation of aeromagnetic maps: pseudo-gravimetric anomalies. Geophysics 1957:.22.359-383

20. Grauch, V. J. S., Cordell, L. Limitations on determining density or magnetic boundaries from horizontal gradient of gravity or pseudogravity data. Geophysics 1987: 52.118-121

21. Grauch, V. J. S. A new variable magnetization terrain correction method for aeromagnetic data. Geophysics 1987: 52 94-107.

22. Nettleton, L. L. Gravity and magnetic calculations. Geophysics 1942:7.293-310

23. Talwani, M., Heirtzler, J. M. Computation of magnetic anomalies caused by two-dimensional structures of arbitrary shape, In: Computers in mineral industries, G. A. Parks (Ed.) 464-480, Stanford Univ.: 1964

24. Talwani, M. Computation with help of a digital computer of magnetic anomalies caused by bodies of arbitrary shape. Geophysics 165: 30 797-817

25. Bhattacharyya, B. K. Continuous spectrum of the total magnetic field anomaly due to a rectangular prismatic body, Geophysics 1966: 31 97-121

26.   Hjelt, S. Magnetostatic anomalies of dipping prisms. Geoexploration 1972:10, 239-254

27.   Werner, S. Interpretation of magnetic anomalies at sheet-like bodies. Sveriges Geol. Undersok. Ser. C, Arsbok, 1953:43(6)

28.   O'Brien, D. P. CompuDepth: a new method for depth-to-basement computation. Presented at the 42nd Annual International Meeting 1972, SEG.

29.   Naudi, H. Automatic determination of depth on aeromagnetic profiles. Geophysics 1971: 36 717-722

30.   Nabighian, M. N. The analytic signal of two-dimensional magnetic bodies with polygonal cross-section: its properties and use for automated anomaly interpretation. Geophysics 1972:.37 507-517

31.   Nabighian, M. N. Additional comments on the analytic signal of two-dimensional magnetic bodies with polygonal cross-section. Geophysics 1974: 39 85-92

32.   Thompson, D. T. EULDPH – a new technique for making computer-assisted depth estimates from magnetic data. Geophysics 1982: 47 31-37.

33.   Thurston, J. B., Smith, R. S. Automatic conversion of magnetic data to depth, dip and susceptibility contrast using the SPI ™ method. Geophysics 1997: 62.807-813

34.   Spector, A., Grant, F. S. Statistical models for interpreting aeromagnetic data. Geophysics 1970: 35 293-302

35.   Cordell, L., McCafferty, A. E. A terracing operator for physical property mapping with potential field data. Geophysics 1989: 54 621-634.

36.   Grant, F. S. The magnetic susceptibility mapping method for interpreting aeromagnetic survey:43rd Annual International Meeting, SEG 1973 expanded abstract 1201.

37.   Pedersen, L. B.Interpretation of potential field data – a generalized inverse approach. Geophysical Prospecting, 1977: 25 199-230.

38.   Pilkington, M. & Crossley, D. J. Inversion of aeromagnetic data for multilayered crustal models. Geophysics 1986: 51 2250-2254.

39.   Pustisek, A. M. Noniterative three-dimensional inversion of magnetic data. Geophysics 1990:55 782-785.

40. Li, Y. & Oldenburg, D. W. 3-D inversion of magnetic data. Geophysics 1996:61 394-408.

41. Shearer, S. & Li, Y. 3 D inversion of magnetic total gradient in the presence of remanent magnetization. 74th Annual International Meeting, SEG 2004: Abstracts 774-777

42. Kellogg, O. D. Foundations of potential field theory. Dover New York; 1953

43. Nabighian, M. N. Toward a three dimensional automatic interpretation of potential field data via generalized Hilbert transforms: fundamental relation,.Geophysics 1984:49 957-966

44. Spector, A. Spectral analysis of aeromagnetic data. Ph.D thesis, University of Toronto; 1968

45. Moreau, F. D. G., Holschneider, M., Saracco, G. Wavelet analysis of potential fields. Inverse Problems 1997:13 165-178

46. Pelton, C. A computer program for hill-shading digital topographic data sets. Computers and Geosciences 1987:13 545-548.

47. Rajagopalan, S., Milligan, P. Image enhancement of aeromagnetic data using automatic gain control. Exploration Geophysics 1995: 25 173-178.

48. Dentith, M., Cowan, D. R., Tompkins, L. A. Enhancement of subtle features in aeromagnetic data. Exploration Geophysics 2000:31 104-108

49. Carter, J. D., Barber, W., Tait, E. A., Jones, G. P. The geology of parts of Adamawa, Bauchi and Bornu provinces in northeastern Nigeria. Bull. Geol. Survey Nigeria 1963:30 p108

50. Cratchley, C. R. & Jones, G. P. An interpretation of the geology and gravity anomalies of the Benue Valley, Nigeria, Overseas Geological Surveys, Geophysical Paper 1965 (1)

51. Burke, K., Dessauvagie, T. F. J., Whiteman, A. J. Geologica history of the Benue Valley and adjacent areas, In: African Geology, T. F. J. Dessauvagie & A. J. Whiteman (Eds.) University of Ibadan Press, Nigeria; 1970

52. Benkhelil, M. J. The origin and evolution of the Cretaceous Benue Trough, Nigeria, Journal of African Earth Sciences 1989: 8 251-282

53. Roest, W. R., Verhoef, J., Pilkington, M. Magnetic interpretation using 3-D analytic signal. Geophysics 1992: 57 116-125

54. Debeglia, N. & Corpel, J. Automatic 3-D interpretation of potential field data using analytic signal derivatives. Geophysics 1997:62 87-96

Simulation of Tsunami Impact on Sea Surface Salinity along Banda Aceh Coastal Waters, Indonesia

Maged Marghany[1]

[1]Institute of Geospatial and Science Technology (INSTEG), University of Technology, Malaysia

INTRODUCTION

Constantly later the catastrophe of the Indian Ocean tsunami of Boxing Day 2004, research in tsunami geoscience has augmented evidently [1]. Scientists have attempted to comprehend the mechanisms of the wide scale of the Indian Ocean tsunami of 2004. Nevertheless, with great efforts done by scientists since Boxing day 2004, the Japanese tsunami with great disaster occurred. However, on March 11, 2011, a magnitude of M_w 9.0 earthquake struck off the coast of Honshu, Japan, sparking a tsunami that not only devastated the island nation, but also caused destruction and fatalities in other parts of the world, including Pacific islands and the United State (U.S.) West Coast [4].

Initial reports were eerily similar to those on December 26, 2004, when a massive underwater earthquake off the coast of Indonesia's Sumatra Island rattled the Earth in its orbit. The 2004 quake, with a magnitude of M_w 9.1, was the largest one since 1964. But as in Japan, the most powerful and destructive aftermath of this massive earthquake was the tsunami that it caused. The death toll reached higher than 220,000 [1-3][10].

Definitions of Tsunami

It is well known that the tsunami is the natural phenomena consisting of a series of waves generated when the waves are rapidly displaced on a massive scale. Tsunami (pronounced soo-NAH-mee) is a Japanese word which is meaning harbor ('tsu") and wave ("nami"). Tsunamis are fairly common in Japan and many thousands of Japanese have been killed by them in recent centuries. In this context, the term was coined by fishermen who returned to port to find the area surrounding the harbor devastated, in spite they had not been aware of any wave on high seas [2].

Haugen et al., [8] stated that tsunamis are long waves set in motion by an impulsive perturbation of the sea, intermediate between tides and swell waves in the spectrum of gravity water waves. Subsequently, Zahibo et al., [5] defined tsunami waves as surface gravity waves that occur in the ocean as the result of large-scale short-term perturbations (underwater earthquakes, eruptions of underwater volcanoes, landslides, rock falls, pyroclastic avalanche from land volcanoes entered in water, asteroid impact, and underwater explosions.

Comments on Tsunami Definition

In earlier times, seismic ocean waves were called "tidal" waves, incorrectly implying that they had some direct connection to the tides. In fact, when the tsunami approach coastal zone they began to characterize by a violent onrushing tidal rather than the sort of cresting waves that are generated by wind stress upon the sea surface. However, to eliminate this confusing the Japanese word "tsunami is used to describe the giant wave (Figure 1) in which is referring to a seismic wave and meaning harbor wave to replace the misleading term tidal wave. This tsunami is a synonym for seismic sea wave. In this regard, a tsunami is a seismic sea wave containing tremendous amounts of energy as a result of its mode of formation i.e. the factor that causing a seismic wave. Therefore, tsunamis are temporary oscillations of sea level with periods longer than wind, waves and shorter than tides for the tsunami, and shorter than a few days of storm surge [8].

Figure 1: Giant Tsunami Wave [3].

Tsunami Characteristics

The physical parameters of duration, length, propagation speed, and heights are the keys description of tsunami. In this regard, tsunamis have duration is ranged between 5 to 100 minutes. They have long length which is ranged between100 m to1000 km. Further, tsunami propagation speed is between 1 to 200 m/s, and their heights can be up to tens of meters. Therefore, Zahibo et al., [5] stated that tsunami waves of the seismic origin are usually very long (50–1000 km). In this context, the source of the giant 2004 tsunami in the Indian Ocean (magnitude of M_w 9.0–9.3) has approximated dimensions: (i) ength; (ii) 670 km; (iii) width 150 km; and (iv) height 12 m [5].

This means that a tsunami can propagate long distances before it strikes a shoreline hundreds or thousands of kilometres from the earthquake source. To accurately model tsunami propagation over such large distances, the Earth's curvature should be taken into account. Other factors, such as Coriolis forces and dispersion, may also be important [7].

Tsunami Generation

It is interesting to understand the mechanisms of tsunami generation. Generation mechanisms of tsunamis are geological events like land- and rockslides in fjords and lakes, submarine gravity mass flows and earthquakes. The Storage slide is one of the largest submarine slides in the world [8]. While the mechanism for generating the initial water waves by purely tectonic motions is reasonably well comprehended. Conversely, the modelling of tsunamis generated by submarine landslides is not yet explicated. Co-seismic deformation of the seafloor usually occurs rapidly relative to the propagation speeds of long water waves (Figure 2), allowing for simple specification of initial conditions by transferring the resultant permanent seafloor deformation to the free surface. However, sub-aerial and submarine landslides move less rapidly and the time-history of seafloor deformation (Figure 2) is important, necessitating the addition of source terms in the equations of motions. Compared with the understanding of earthquake-induced initial tsunami waves, the understanding of landslide-generated waves is marginal. Briefly, in terms of the semi-analytical empirical studies transferred the energy released by a moving block sliding from its initial position to its final position to solitary waves and calculated the height of the wave [9].

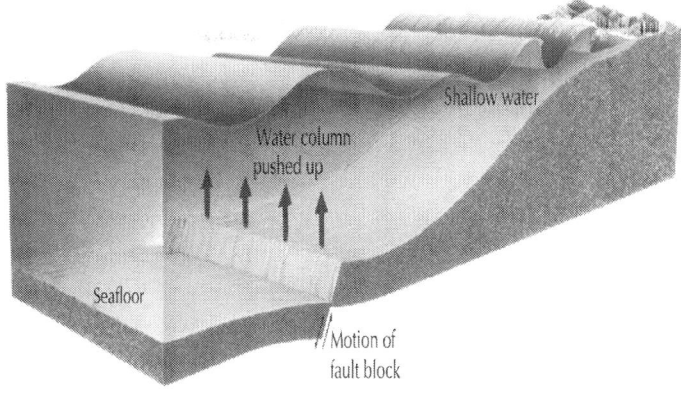

Figure 2: Tsunami generation due to motion of fault block.

Tsunami Classifications

Tsunami can be classified by the distance from their source to the area of impact; i.e., local and remote tsunami. Locally generated tsunami have short warning times and relatively short wave periods; remote tsunami have longer warning times and relatively long periods. Typical periods for tsunami range from 15 minutes for locally generated tsunami to several hours for remote tsunami. Typical run-up height for tsunami range up to 15 m at the coast, although most are much smaller. Storm surges on the other hand are caused by variations in barometric pressure and wind stress over the ocean. Decreasing barometric pressure causes an inverse barometer effect where sea level rises. This is usually a slow and large-scale effect and thus does not usually generate waves in the frequency range typical of tsunami. However, there can be short-period meteorological events (such as meteorological tsunami – rissaga) with time-scales of a few hours that may be important. Wind stress on the other hand has a wide range of time-scales and causes coastal sea-level setup as well as wind waves, where the setup depends on the wind direction, strength, and wave height. Storm surge periods range from several hours to several days. Typical heights for storm surge alone range up to 1.0 m for instance, along the coast of New Zealand, although most are usually less than 0.5 m. Wind waves, on the other hand, can be quite large, producing wave setup and wave run-up of several meters in height.

In general, tsunamis can be categorized into: (i) microtsunami which has a small amplitude and would not even notice visually and (ii) local tsunami which has destructive impact due to its wide spread on coastal zone within hundred of kilometres and usually caused by plate tectonic movements. It might be internal wave named by internal tsunami due to its traveling along the a thermocline layer.

In a somewhat similar fashion, dropping a stone into a puddle of water creates a series of waves which radiate away from the impact point. In this context, the impact point of puddle of water is representing a sudden shifting of rocks or sediments on the ocean floor caused by cataclysmic event, such as a volcanic eruption, an earthquake, or a submarine landslide, can force the water level to drop ≥ 1 m generating a tsunami-a series of low waves with long periods, and long wavelengths (Figure 3). This indicates that the tsunami wavelength is bigger than its amplitude in the open ocean. Thus the tsunami height in the open ocean is approximately less than 1 m which it is not noticeable in open ocean. The tsunamis, however, cross the oceans at a rate of ~750 km/hr which is equivalent to a speed of a modern jet aircraft! Despite these phenomenal speeds, tsunami pose no danger to vessels in the open ocean. Indeed, the regular ocean swell would probably mask the presence of these low sea waves. The tsunamis, however, grow to height of ≥ 10 m as it impinges on a shoreline and flood the coast, sometimes with catastrophic results, including widespread property damage and loss of life.

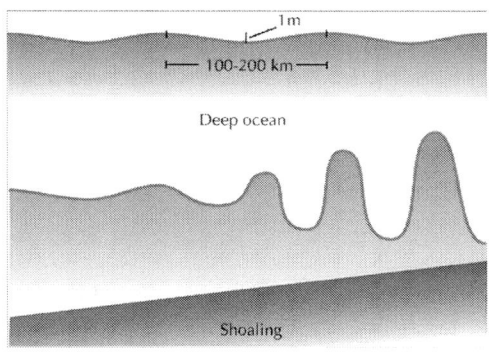

Figure 3: Tsunami Generation in Deep Water.

Geological Descriptions of Sumatra Earthquake

In previous sections, the fundamental mechanism of tsunami was explained. Therefore, this section is devoted for tsunami of 26 December 2004 which is known as Sumatra-Andaman earthquake. Consequently, this disaster is called as Asian Tsunami in Asia region and also known as a Boxing day in the Australia, Canada, New Zealand, and the United Kingdom, because it took place on Boxing Day. This earthquake was also reported to be the longest duration of faulting ever observed, lasting between 500 and 600 seconds (8.3 to 10 minutes), and it was large enough that it caused the entire planet to vibrate as much as half an inch, or over a centimeter.

Epicenter of the Giant Tsunami

The epicentre of the earthquake of the exceptionally high magnitude of 9.0 (Figure 4) is situated inside the trough as indicated between the northern edge of Sumatra and the small island of Simeulue, one member of the chain of islands next to the trench. Neither this trough relief reveals anything unusual nor does the comparatively moderate depth of the Sunda Trench of less than 5, 000 m, the shallower sister of the conspicuous, adjacent Java-Trench southeast of it (Figure 5). The existing records of earthquakes of more than magnitude 5.5 in the Sumatra area since 1970 have clearly shown that their great majority, including the epicentre of the Giant Tsunami, occurred within the shallow strip between the coastline of Sumatra and the adjacent subduction trench rim; in other words: they occurred on the edge of the overriding plate!

Figure 4: Epicenter of Giant Tsunami.

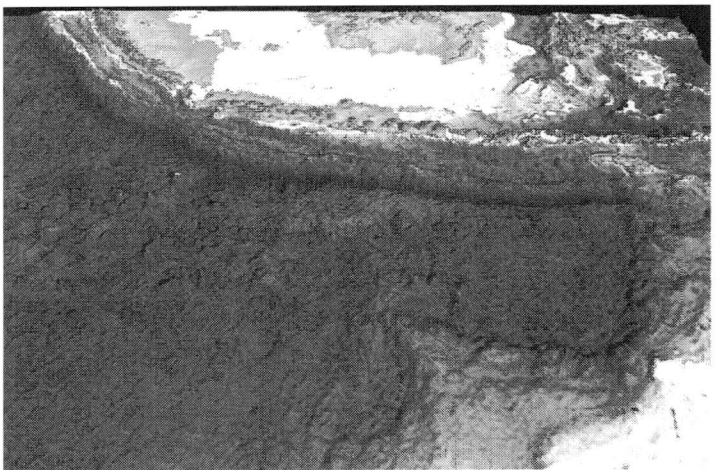

Figure 5: Java Trench.

According to Abbott [11], the Indian-Australian plate moves obliquely toward western Indonesia at 5.3 to 5.9 cm/yr (2 to 2.3 in/yr). The enormous, ongoing collision results in subduction-caused (Figure 6) earthquakes that are frequent and huge. Many of these earthquakes send off tsunami. On 26 December 2004, 1.200 km (740 mi) long fault rupture began as 100 km (62 mi) long portion of the plate tectonic boundary ruptured and slipped during a minute (Figure 6). The rupture then moved northward at 3 km/sec (6,700 mph) for four minutes. Then slowed to 2.5 km/sec (5,6000) during the next six minutes. At the northern end of the rupture, the fault movement slowed drastically and only traveled tens of meters during the next hour. It appears that a bend or scissors like tear in the subduction plate may have delayed the full rupture in December 2004.

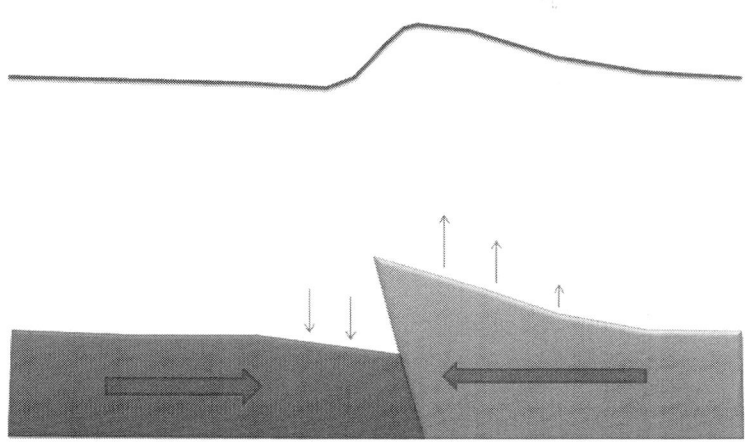

Figure 6: Deformation of sea Flooar and Sea Surface.

Shulgin et al., [13] stated that the tectonics around Northern Sumatra are predominantly controlled by the subduction of the oceanic Indo-Australian plate underneath Eurasia. The current convergence rate offshore North-ern Sumatra is estimated at 51 mm/yr. The increasing obliquity of the convergence northwards from the Sunda Strait results in the formation and development of a number of arc-parallel strike-slip fault systems (Figure 7). The most significant are the Sumatra and the West Andaman Fault systems, accom-modating arc-parallel strain offshore central-southern Sumatra. For the Mentawai fault system. recent findings suggest deforma-tion dominated by backthrusting.

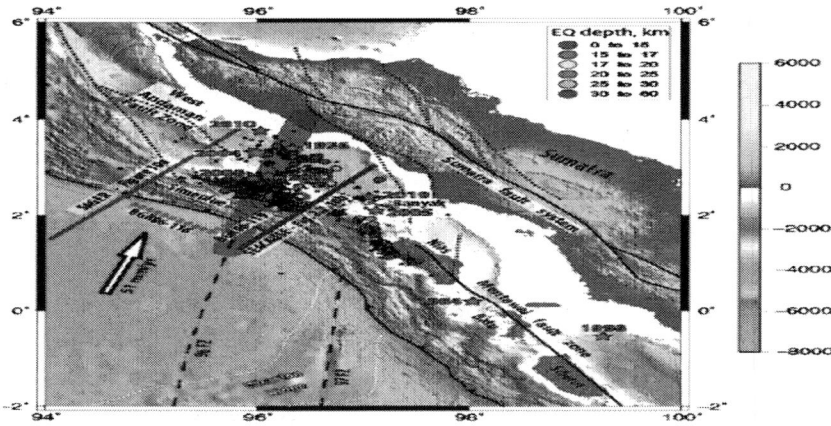

Figure 7: Subduction of Indian-Australian plate beneath Indonesia Shulgin et al., [13].

Subsequently, Kelmelis et al., [10], the Indo-Australian plate is moving approximately 40 to 50 mm per year to the northeast, and the earthquake ruptured at the subduction zone boundary or interpolate thrust boundary (Figure 8). On the fault the earthquake had a maximum slip of approximately 15 to 20 meters with an average slip of >5 meters along the full length of the rupture. The sea floor overlying the thrust fault would have been raised by several meters. Further, velocities of displacement along 1200 to 1300 kilometres of the fault with at least three major energy bursts during the propagation of the rupture (50 to 150 seconds, 280 to 340 seconds, and 450 to 500 seconds.

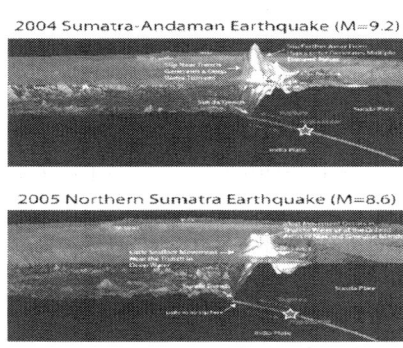

Figure 8: Location of Hypocenter of 2004.

2004 Tsunami Wave Propagation

At 00:58:53 UTC (07:58:53 local time) December 26, 2004, an undersea earthquake was occurred along a tectonic subduction zone in which the India Plate. In general Figure 9 shows that the tsunami reaches Phuket and Sri Lanka coasts in two hours after the earthquake, and African coast in 8-11 hours. The tsunami propagation is also animated (up to 5 hours) from a 1200 km fault. The red color means that the water surface is higher than normal, while the blue means lower. It indicates that initial tsunami to the east (e.g., Phuket) began with receding wave, while to the west (e.g., Sri Lanka) large wave suddenly reached. The darker the color, the larger the amplitude. The tsunamis were larger in the east and west directions.

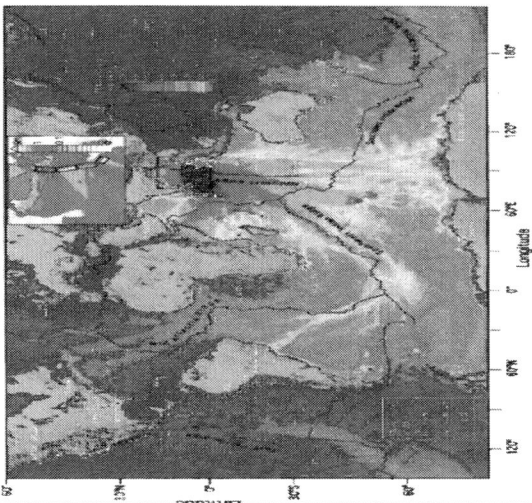

Figure 9: Tsunami wave propagation during 2004 [14].

The tsunami was caused by movement along a fault line running through the seabed of the Indian Ocean. As the fault runs north-south, the waves Travelled out across the ocean in mainly easterly and westerly direction with duration of 7 hours that shook the world. At 00.58 GMT, an undersea earthquake measuring 9.3 on the Richter Scale occurs off Indonesia. At+15 minutes later, the Indonesia Island of Sumatra, close to epicenter of the quake, is hit by the full force of tsunami. Many towns and villages in Aceh province on the western tip of the island are completely washed away, and the capital, Banda Aceh, is destroyed.

The remote Andaman which is lying only 100 km from the epicenter of the earthquake was struck within+30 minutes later. An hour after it hit Sumatra, the tsunami reached Thailand. It had lessened slightly in height and power but still struck the Thai coast with incredible force and the sea surged out for about 200 m. One of the worst-hit places was Sri-Lanka, which lay almost directly west of the earthquake's epicenter. The tsunami wave reached Sri Lanka within+2.00 hours later. The tsunami height recorded in Sri Lanka ranged between 5 to 10 m (Table 1).

Table 1: Measured Tsunami Wave Heights on Boxing Day

Location	Heights
Sri Lanka, east coast	5-10 m
India, east coast	5-6 m
Andaman Islands	>5 m
Phuket, Thailand	3-5 m
Kenya	2-3 m

With no continental shelf to lift the tsunami's waves as they near shore, the Maldives gets off relatively lightly within+3.30 hours. Finally tsunami waves strike Africa coast within+7 hours. The wave height along Kenya coast was between 2 and 3 m (Table 1).

Further, the Christmas tsunami was so powerful it actually sped up the rotation of the Earth reducing the length of its sidereal day. The earthquake that spawned it also caused the Earth to vibrate all over by as much as 1 cm. In this regard, the critical question may be raised is what the tsunami effects on the ocean physical properties such as temperature and salinity? In fact, there are no many studies have been conducted to answer this question. Temperature and salinity are the main parameters can be used to describe the ocean status. Both parameters can produce vertical current movement because of the their gradient changes. In addition, water density changes are function of gradual changes of temperature and salinity.

Hypotheses and Objective

Concern with above prospective, we address the question of tsunami 's impact on Sea Surface Salinity (SSS) pattern changes pro and post tsunami event of 2004. This is demonstrated with Moderate-resolution Imaging Spectrometer (MODIS) i.e. the Aqua/MODIS data level IB reflectance satellite data using least square algorithm. Three hypotheses are examined:

- Least square algorithm can be used to retrieve Sea Surface Salinity in MODIS satellite data during and post tsunami disaster;
- Least square algorithm is automatic algorithm for retrieving SSS for short periods i.e. couple of days;
- There is changes in SSS pattern post tsunami 2004 event along Banda Aceh coastline.

STUDY AREA

The study area is located along the western coastal zone of Aceh with boundaries of latitudes 3° 30´ N to 6° 30´ N and longitudes of 93° 30´ E to 99° 30´E (Figure 10). The Sunda trench is running north-south along the coastal waters of Aceh towards the Andaman Sea. Running in a rough north-south line on the seabed of the Andaman Sea is the boundary between two tectonic plates, the Burma plate and the Sunda Plate. These plates (or microplates) are believed to have formerly been part of the larger Eurasian Plate, but were formed when transform fault activity intensified as the Indian Plate began its substantive collision with the Eurasiancontinent. As a result, a back-arc basin center was created, which began to form the marginalbasin which would become the Andaman Sea, the current stages of which commenced approximately 3–4 million years ago (Figure 11). On December 26, 2004, a large portion of the boundary between the Burma Plate and the Indo-Australian Plate slipped, causing the 2004 Indian Ocean earthquake. This megathrust earthquake had a magnitude of 9.3. Between 1300 and 1600 kilometers of the boundary underwent thrust faulting and shifted by about 20 meters, with the sea floor being uplifted several meters. This rise in the sea floor generated a massive tsunami with an estimated height of 28 meters (30 ft) [14].

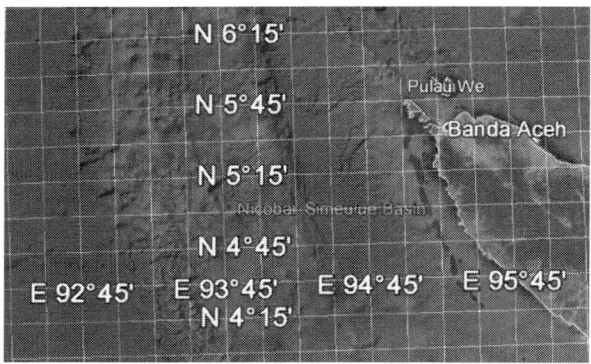

Figure 10: Location of study area.

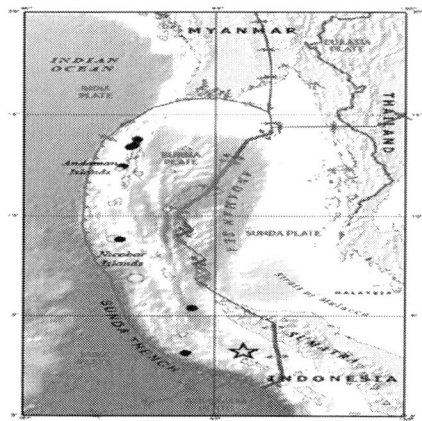

Figure 11: Location of Sunda Trench.

The average depth of the sea is about 1,000 meters (3,300 ft). The northern and eastern parts are shallower than 180 meters (600 ft) due to the silt deposited by the Irrawaddy River. This major river flows into the sea from the north through Burma. The western and central areas are 900–3,000 meters deep (3,000–10,000 ft). Less than 5% of the sea is deeper than 3,000 meters (10,000 ft), and in a system of submarine valleys east of the Andaman-Nicobar Ridge, the depth exceeds 4,000 meters (13,200 ft). The sea floor is covered with pebbles, gravel and sand [14].

Further, the climate and water salinity of the Andaman Sea and Aceh are mostly determined by the monsoons of southeast Asia. Air temperature is stable over the year at 26 °C in February and 27 °C in August. Precipitation is as high as 3,000 mm/year and mostly occurs in summer. Sea currents are south-easterly and easterly in winter and south-westerly and westerly in summer. The average surface water temperature is 26–28 °C in February and 29 °C in May. The water temperature is constant at 4.8 °C at the depths of 1,600 m and below. Salinity is 31.5–32.5‰ (parts per thousand) in summer and 30.0–33.0‰ in winter in the southern part. In the northern part, it decreases to 20–25‰ due to the inflow of fresh water from the Irrawaddy River. Tides are semidiurnal (i.e. rising twice a day) with the amplitude of up to 7.2 m [23].

LEAST SQUARE MODEL

In this section, we present the theoretical model of split window method that relates MODIS sea surface salinity with in situ salinity measured by thermal infrared sensors, these include multi-channel methods. We assume the MODIS image radiance

Within multi-channels i have a linear relationship with measured sea surface salinity (SSS). A useful extension of linear function of k channels as in

$$SSS = b_0 + b_1 I_1 + b_2 I_2 + b_3 I_3 + \ldots\ldots\ldots + b_k I_k + \varepsilon \tag{1}$$

The model may be written in terms of the observations as

$$SSS = b_0 + \sum_{i=1}^{k} b_i I_i + \varepsilon \tag{2}$$

where SSS is the measured sea surface salinity, k is a number of MODIS radiance bands which equals 7bands, b_0, b_i are constant coefficient of linear relationship between MODIS radiance data (I) and ε is residual error of SSS estimated from selected MODIS bands. The unknown parameters in equation 2, that are b_0 and b_i may be estimated by a general least square iterative algorithm. This procedures requires certain assumptions about the model error component ε. Stated simply, we assume that the errors are uncorrelated and their variance is σ_ε^2. In general, if ε_i and ε_j are two uncorrelated errors, then their covariance is zero, where we define the covariance as

$$Cov(\varepsilon_i, \varepsilon_j) = E(\varepsilon_i \cdot \varepsilon_j) \tag{3}$$

The least-square estimator of the b_i minimizes the sum of squares of the errors, say

$$= \sum_{j=1}^{n}(SSS_j - b_0 - \sum_{i=1}^{k}b_i I_i)^2 = \sum_{j=}^{n} \tag{4}$$

where SSS_j is the value of SSS measured at I_i, n is the total number of data points and $n \geq k$. It is necessary that the least squares estimators

satisfy the equations given by the k first partial derivatives $\frac{\partial S_E}{\partial b_i} = 0$, $i=1,2,3,.....,k$ and $j=1,2,3,.....,n$. Therefore, differentiating equation 4 with respect to bi and equating the result to zero we obtain

$$n\hat{b}_1 + (\sum_{j=1}^{n}I_{2j})\hat{b}_2 + (\sum_{j=1}^{n}I_{3j})\hat{b}_3 + ... + (\sum_{j=1}^{n}I_{kj})\hat{b}_k = \sum_{j=1}^{n}SSS_j$$

$$(\sum_{j=1}^{n}I_{2j})\hat{b}_1 + (\sum_{j=1}^{n}I^2{}_{2j})\hat{b}_2 + ... + (\sum_{j=1}^{n}I_{2j}I_{kj})\hat{b}_k = \sum_{j=1}^{n}I_{2j}SSS_j$$

$$(\sum_{j=1}^{n}I_{kj})\hat{b}_1 + (\sum_{j=1}^{n}I_{kj}I_{2j})\hat{b}_2 + ... + (\sum_{j=1}^{n}I^2_{kjj})\hat{b}_k = \sum_{j=1}^{n}I_{kj}SSS_j$$

$$\tag{5}$$

The equations (5) are called the least-squares normal equations.

The $\overrightarrow{b_k}$ found by solving the normalequations (5) are the least-squares estimators of the parameters b_i. The only convenient way to express the solution to the normal equations is in matrix notation. Note that the normal equations (5) are just a $k \times k$ set of simultaneous linear equations in k unknowns (the $\{ \overrightarrow{b_k} \}$). They may be written in matrix notation as

$$H\hat{b} = h \tag{6}$$

Where

$$H = \begin{bmatrix} n & \sum I_{2j} & \cdots\cdots & \sum I_{kj} \\ \sum I_{2j} & \sum I_{2j}^2 & \cdots\cdots & \sum I_{2j}I_{kj} \\ \sum I_{3j} & \sum I_{3j}I_{2j} & \cdots\cdots & \sum I_{3j}I_{kj} \\ \cdots\cdots & \cdots\cdots & \cdots\cdots & \cdots\cdots \\ \sum I_{kj} & \sum I_{kj}I_{2j} & \cdots\cdots & \sum I_{kj}^2 \end{bmatrix}$$

$$\hat{b} = \begin{bmatrix} \hat{b_1} \\ \hat{b_2} \\ \vdots \\ \vdots \\ \hat{b_k} \end{bmatrix} \text{ and } h = \begin{bmatrix} \sum SSS_j \\ \sum I_{2j}SSS_j \\ \vdots \\ \sum I_{kj}SSS_j \end{bmatrix}$$

Thus, H is a $k \times k$ estimated matrix of MODIS radiance bands that used to estimate sea surface salinity, $\hat{b}$ and h are both $k \times 1$ column vectors. The solution to the least-squares normal equation is

$$\hat{b} = H^{-1}h$$

(7)

where H^{-1} denotes the inverse of the matrix H. Given a solution to least squares normal equations, the retrieval SSS$_{MODIS}$ is estimated using the fitted multiple regression model of equation 2 as given

$$SSS_{MODIS} = \hat{b_1} + \sum_{i=1}^{k} \hat{b_i}I_i$$

(8)

Following Sonia et al., (2007), errors that represents the difference between retrieved and in situ SSS are computed within 10 km grid point interval and then averaged over all grid points having the same range of distance to coast, where the bias on the retrieved SSS$_{MODIS}$ is given by:

$$\varepsilon = \frac{\sum_{i=1}^{N}(SSS_{MODIS} - SSS_{situ})}{N} \tag{9}$$

where SSS_{MODIS} is the retrieved sea surface salinity from MODIS satellite data, SSS_{situ} is the reference sea surface salinity on grid point i and N is the number of grid points. Then, the empirical formula of SSS_{MODIS} (psu) which is based on equations 8 and 9 is

$$) = 27.40 + 2.0I_1 - 3.4I_2 + 2.0I_3 + 2.2I_4 + 1.8I_5 + 0.3I_6 + \tag{10}$$

This equation is different than the equation was obtained by Wong et al. [21] in terms of constant coefficient of linear model and involving the retrieved SSS_{MODIS} bias value. Finally, root mean square of bias (RMS) is used to determine the level of algorithm accuracy by comparing with in situ sea surface salinity. Further, linear regression model used to investigate the level of linearity of sea surface salinity estimation from MODIS data. The root mean square of bias equals [17]

$$RMS = [N^{-1}\sum_{i=1}^{N}(SSS_{MODIS} - SSS_{situ})^2]^{0.5} \tag{11}$$

RESULTS AND DISCUSSION

The tsunami impact on sea surface salinity has been examined on three MODIS satellite data along Aceh coastal waters. These data are acquired on 23rd, 26th and 27th 2004 which represent pro, and post tsunami event, respectively (Figure 12). According to Marghany et al., [18], Moderate Resolution Imaging Spectroradiometer (MODIS) has 1 km spatial resolution and having 36 bands which are ranged from 0.405 to 14.285μm [24]. The MODIS satellite takes 1 to 2 days to capture all the scenes in the entire world, acquiring data in 36 spectral bands over a 2330 km swath.

(a)   (b)

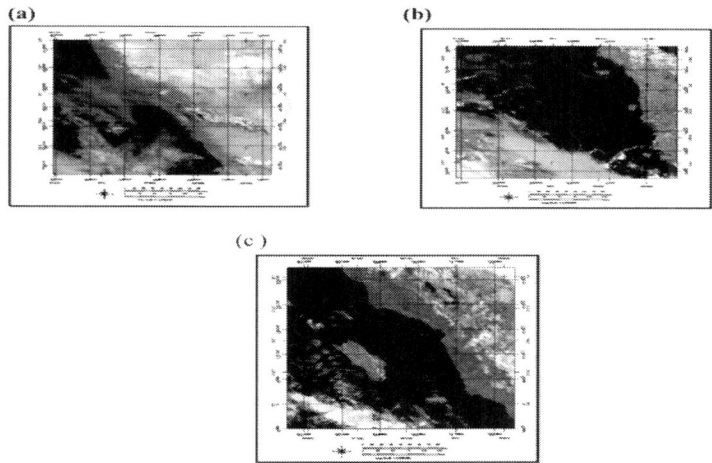

(c)

Figure 12: MODIS satellite data (a) pro tsunami event, (b) during Tsunami event and (c) post tsunami event.

It is interesting to find that heavy cloud covers have occurred on December 23th 2004 as compared to post tsunami event on December 27th 2004 (Figure 12). In fact, Aceh is located on tropical zone where the heavy cloud covers are one of the dominant features. Figure 9 shows the spatial variation of the salinity distribution along Aceh which are derived using the linear least square algorithm. On December 23th 2004, the sea surface salinity ranged between 28 psu to 31 psu. Nevertheless, during the tsunami event 25th 2004, the sea surface salinity dramatically increased and ranged between 34 psu to 36 psu. The sea surface salinity was continued to increase post tsunami event of December 27th 2004 and ranged between 37 psu to 38 psu.

The isohaline contours of sea surface salinity which derived from in situ data. These data were acquired from http://aquarius.nasa.gov/. Figure 13 shows that the in situ sea surface salinity were increased dramatically. Pre tsunami event, the isohaline contours were ranged between 28.5 to 29.0 psu. However, the isohaline contours were increased to 36.7 psu during tsunami event and continued to increase to 37 psu (Figure 13).

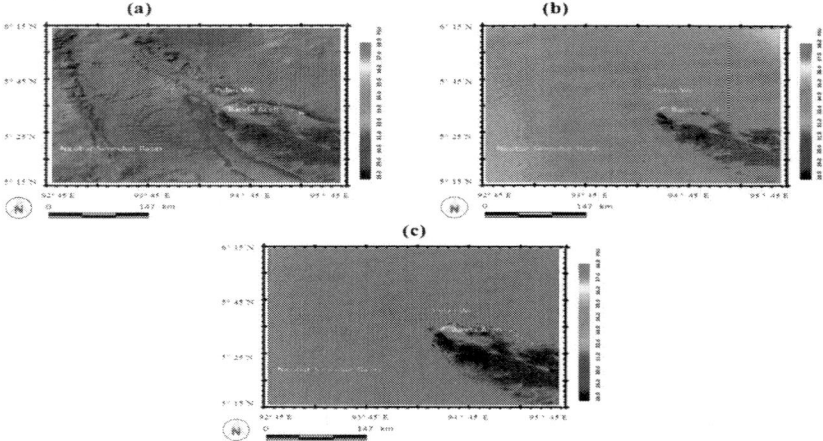

Figure 13: Isohaline contours (a) pro tsunami event, (b) during tsunami event and (c) post tsunami event.

Both MODIS satellite data and in situ data show homogenous sea surface salinity pre tsunami and post tsunami with dramatically increment of sea surface salinity from 28.5 psu to 38.0 psu in coastal waters of Aceh. Figure 15 shows the comparison between in situ sea surface salinity measurements and SSS modeled from MODIS data. Regression model shows that SSS modeled by using linear least square methods are in good agreement with in situ data measurements. The degree of correlation is a function of r^2, probability (p) and root mean square of bias (RMSE). The relationship between estimated SSS from MODIS data and in situ data shows positive correlation as r^2 value is 0.96 with $p<0.00007$ and RMS value of $\pm$ 1.1 psu. Further, accurate results of sea surface salinity in recent study can be explained as: using multiple MODIS bands i.e. 1 to 7 bands is a useful extension of linear regression model is the case where SSS is linear function of 7 independent bands. Such a practical is particularly useful when modeling SSS from MODIS data. This statement is agreed with Qing et al., [24].

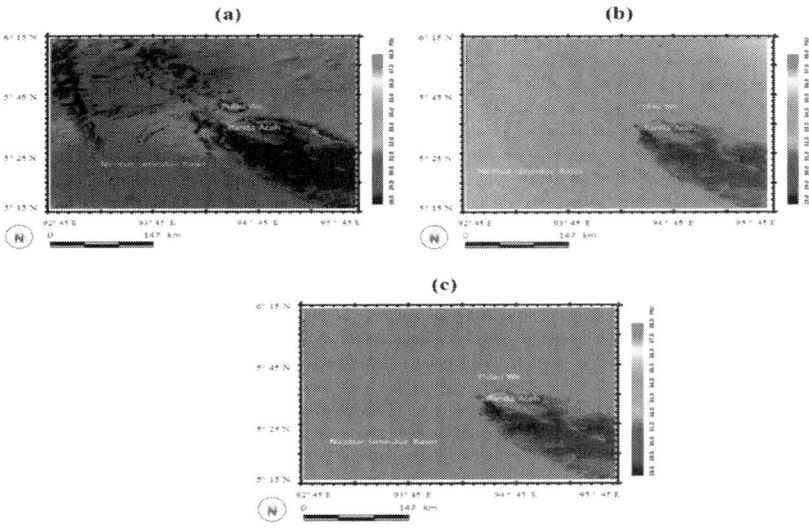

Figure 14: Sea Surface Salinity (SSS) retrieved from MODIS data (a) pro tsunami event, (b) during Tsunami event and (c) post tsunami event.

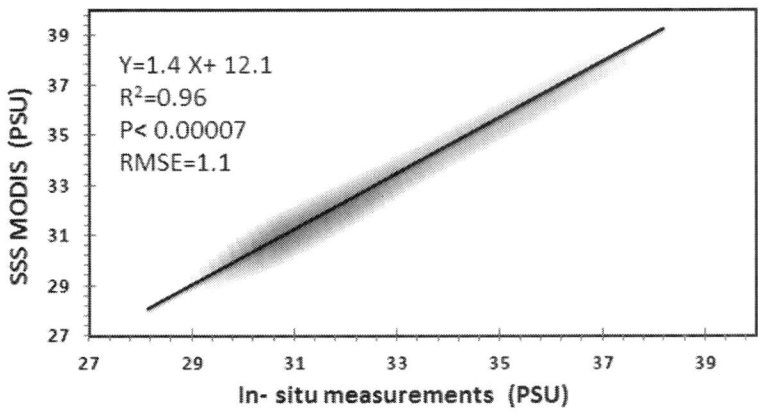

Figure 15: Regression model between in-situ measurement and modeled SSS from MODIS data.

Further, using least squares method derive a curve that minimizes the discrepancy between estimated SSS from MODIS data and in situ data [16]. This means that using a new approach based on least squares method would be to minimize the sum of the residual errors for the estimating SSS from MODIS data. Further, this study shows the possibilities of direct retrieving of the SSS from visual bands of MODIS satellite data without utilizing such parameter of colored dissolved organic matter, a_{CDOM} [15]. This study confirms the studies done by Marghany et al. [18]; Marghany and Mazlan [19].

Further dissolved salts, suspended substances have a major impacts on the electromagnetic radiation attenuation outside the visible spectra range [21]. In this context, the electromagnetic wavelength larger than 700 nm is increasingly absorbed whereas the wavelength less than 300 nm is scattered by non-absorbing particles such as zooplankton, suspended sediments and dissolved salts Marghany et al., [17].

Effect of Changes the Value of Salinity towards the Ocean

The Sea Surface Salinity (SSS) along Banda Aceh has been changed dramatically pro and post tsunami event of 2004. During earlier run-up phases, the flows were be extremely strong to erode sediments deposited especially when the coastal topography and bathymetry canalized the tsunami flow. In this regard, the backwash flows were potentially more erosive and powerful than run-up flows because of hyper-concentrated flow routing by coastal morphology. Therefore, tsunami resulted in massive deposits of sand, silt and fine gravel containing isolated boulders [25]. According to Font et al., [22], the area close to the shoreline was eroded by the passing wave front and deposition occurred when the wave front passed by. Just 6 min later, the direction of the flow close to the shoreline began to change to the seaward direction before the wave front reached the inundation limit. A portion of the sediment entrained by the wave front. In other words, tsunami of 2004 was caused brief coastal flooding with high overland flow velocities and strong abrasion and reworking of the nearshore materials (Figure 16). As most nearshore environments are composed by sand, mud and gravel. In this regard, the salinity values are extremely increased due to tremendous genetic sediment differences carried by wave from inland. According to Moore et al., [23] geochemical proxies have been provided evidence for saltwater inundation, associated coral and/or shell material, high-energy flows (heavy minerals, if present), and possible contamination associated with tsunamis. Finally, Pre and post event imagery can show the extent of erosion and areas most optimal for preservation of tsunami deposits (Figure 17). This information is necessary for comparing different events in the same locality. In addition, In the 2004 tsunami, sediments varied in their salinity levels, so sediments need to be assessed for salinity before any crops are planted in them.

Figure 16: High deposits flow due tsunami 2004 along Aceh coastal waters.

Figure 17: Satellite data along Aceh coastal waters (a) pro and (b) post tsunami event 2004 event.

In general, different grain size of sediments deposited in the coastal waters of Banda Aceh have increased the salinity in the sea surface. Indeed, these sediments contain different level of salts and mineral concentration. Post-tsunami, SSS has increased extremely because the sea coast was rough and turbid with suspended sediments (Figures 16 and 17). This is excellent evident of additional of extremely high amount of salts and minerals due high level of sediment deposit concentration in the coastal waters of Banda Aceh. This study confirms the work done by Saraf et al., [26].

CONCLUSIONS

Remote sensing technology has been recognized as powerful tool for environmental disaster studies. Ocean surface salinity is considered as a major element in the marine environment. In this study, we simulated the tsunami 2004 impacts on the Sea Surface Salinity along Banda Aceh using the least square algorithm from MODIS satellite data. This study shows significant variations in the values of SSS pro, during and post the tsunami event. The maximum salinity was observed post tsunami event was 38 psu as compared to before and during tsunami event. The results also show a good correlation between in situ SSS measurements and the SSS that is retrieved from MODIS satellite data with high $r^2$of 0.96 and RMSE of bias value of ±1.1 psu.In conclusion, the least square algorithm is an appropriate method to retrieve SSS from MODIS satellite data. Clearly, the tsunami 2004 has significant impacts on the SSS because of high sediment deposit concentrations which added more salts and minerals to coastal waters of Banda Aceh.

REFERENCES

1.   Dawson A, Stewart I, Tsunami Geoscience Progress in Physical Geography. 2007, 31(6) 575– 590.

2.   University of Washington. Earth & Space science: The Physics of Tsunamis. http://earthweb.ess.washington.edu/tsunami/general/physics/physics.html(accessed 11 November 2013).

3.   Taru T, Japan Experts Warn of Future Risk of Giant Tsunami. http://www.cosmostv.org/2012/04/japan-experts-warn-of-future-risk-of.html. (accessed 11 November 2013).

4. Valdes R, Halabrin N, Lamb R How Tsunamis Work. http://science.howstuffworks.com/nature/natural-disasters/tsunami.htm. (accessed 11 November 2013).

5. Zahibo N, Pelinovsky E, Talipova T, Kozelkov A, Kurkin A. Analytical and numerical study of nonlinear effects at tsunami modelling. Applied Mathematics and Computation. 2006 (174) 795– 809.

6. Zaitsev A, Kurkin A, Levin B, Pelinovsky E, Yalciner A C, Troitskaya Y, Ermakov S. Modeling of propagation of the catastrophic tsunami (December 26, 2004) in the Indian Ocean, Doklady Earth Sci. 403 (3) (2005).

7. V V and Synolakis C E. Numerical modeling of 3-D long wave runup using VTCS-3. In *Long Wave Runup Models*, P. Liu, H. Yeh, and C. Synolakis (eds.), World Scientific Publishing Co. Pte. Ltd., Singapore, 1996 242–248.

8. Haugen K B, Løvholt F, Harbitz C B. Fundamental mechanisms for tsunami generation by submarine mass flows in idealised geometries. Marine and Petroleum Geology. 2005 (22) 209– 217.

9. Liu P L F, Wu T R, Raichlen F, Synolakis C. and Borrero J. Runup and rundown generated by three-dimensional sliding masses. Journal Fluid Mech. (2005) (536),107–144.

10. Kelmelis JA, Schwartz L, Christian C, Crawford M, and King D. Use of Geographic Information in Response to the Sumatra-Andaman Earthquake and Indian Ocean Tsunami of December 26, 2004. Phototogrammetric Engineering & Remote Sensing. 2006 72 (8) 862-876.

11. Abbott P L. Natural Disasters. 6th edition, 2008, McGraw Hill Higher Education, New York.

12. Catherine J K, Gahalaut V K, Ambikapathy A, Kundu B, C. Subrahmanyam, S. Jade, Amit Bansal, R.K. Chadha, M. Narsaiah, L. Premkishore, D.C. Gupta. 2008 Little Andaman aftershock: Genetic linkages with the subducting 90°E ridge and 2004 Sumatra–Andaman earthquake. Tectono physics 2009 (479) 271–276.

13. Shulgin A, Kopp H, Klaeschen D, Papenberg C, Tilmann F, Flueh E R, Franke D, Barckhausen U, Krabbenhoeft A, Djajadihardja Y. Subduction system variability across the segment boundary of the

2004/2005 Sumatra megathrust earthquakes. Earth and Planetary Science Letters 365 (2013) 108–119.

14. Titov V V, Gonzalez F I, Bernard E N, et al. (2005): In: Real-time tsunami forecasting: challenges and solutions. Nat. Hazards 35(1), Special Issue, pp. 41–58. US National Tsunami Hazard Mitigation Program.

15. Ahn Y H, Shanmugam P, Moon J E, and Ryu J H. Satellite remote sensing of a low-salinity water plume in the East China Sea. Annals of Geophys. 2008 (26) 2019–2035.

16. Marghany M. Linear algorithm for salinity distribution modelling from MODIS data, Geoscience and Remote Sensing Symposium,2009 IEEE International, IGARSS 2009, 12-17 July 2009, Cape Town, South Africa, 2009 (3) III-365-III-368.

17. Marghany M. Examining the Least Square Method to Retrieve Sea Surface Salinity from MODIS Satellite Data European Journal of Science of Research. 40 (2010) 377-386.

18. Marghany M Hashim M and Cracknell A P. Modelling Sea Surface Salinity from MODIS Satellite Data. Computational Science and Its Applications – ICCSA 2010, Lecture Notes in Computer Science. 2010, (6016) 545-556.

19. Marghany M and Hashim M. A numerical method for retrieving sea surface salinity from MODIS satellite data. International Journal of the Physical Sciences. 2011 6(13) 3116-3125.

20. Hu C, Chen T, Clayton P, Swarnzenski J, Brock I, and Muller-Karger F. Assessment of estuarine water-quality indicators using MODIS medium-resolution bands: Initial results from Tampa Bay, fL. Remote Sensing of Environment 2004 (93) 423-441

21. Wong M S, Kwan L, Young J K, Nichol J, Zhangging L, Emerson N Modelling of suspendid solids and sea surface salinity in Hong Kong using Aqua/ MODIS satellite images. Korian Journal of Remote sensing. 2007 (23) 161-169.

22. Font E, Nascimento C, Omira R, Baptista MA, Silva PA. Identification of tsunami-induced deposits using numerical modelling and rock magnetism techniques: A study case of the 1755 Lisbon tsunami in Algarve, Portugal. Physics of the Earth and Planetary Interiors. 2010 (182) 187–198.

23. Moore A, Nishimura Y, Gelfenbaum G, Kamataki T, and Triyono R, Sedimentary deposits of the 26 December 2004 tsunami on the northwest coast of Aceh, Indonesia. Earth Planets Space, 2006 (58) 253–258.

24. Qing S Zhang J Cui T Bao Y. Retrieval of sea surface salinity with MERIS and MODIS data in the Bohai Sea. Remote Sensing of Environment. 2013 (136) 117-125.

25. Anthony E J. Developments in marine geology: Shore Processes and their palaeoenvironmental applications. Series Editor Chamley H, 2009 (4) 415-420.

26. Saraf A K, Choundhury and Dasgupta S. Satellite observation of the great mega thrust Sumatra earthquake activities. International journal of Geoinformatics. 2005, 1(4) 67-74.

Chapter 6

HF Radar Network Design for Remote Sensing of the South China Sea

S. J. Anderson[1]

[1]Defence Science and Technology Organisation, Edinburgh SA, Australia

INTRODUCTION

HF surface wave radar (HFSWR) is a highly cost-effective technology for remote sensing of ocean surface conditions and monitoring of ship traffic; several hundred radars of this type are in operation around the world. While an individual radar, operating alone, is able to provide a great deal of useful information, the integration of multiple radars into a network results in a system capability which is far more than the sum of its parts. For example, an estimate of a ship's velocity vector can be obtained in seconds, not tens of minutes or hours as is the case with a single radar. As another example, ocean currents can be estimated unambiguously, even in the presence of eddies and upwelling. Apart from these well-known considerations, there is a class of benefits which has special significance for long range HFSWR systems, namely, the potential for bistatic operations. As shown later in this paper, the fusion of monostatic and bistatic measurements enhances radar performance in a number of ways, a gain which is especially important for very long range operations.

While some HFSWR systems have been designed and deployed with a single mission in mind, it is increasingly recognised that the versatility of this technology supports a variety of applications. For instance, one might wish to detect and track shipping but also to measure surface currents so that risks of collision or grounding can be minimised and any transport of pollution predicted. In addition, information on sea state is of considerable economic value for ship routing, planning for offshore wave energy extraction, coastal development, port operation scheduling, search and rescue, fishing, tourism and recreational activities, so extraction and dissemination of environmental data would be welcomed by a wide range of user communities. Of course, these various applications will have relative priorities which vary with location, time of day and season, as will the radar's ability to accomplish them.

The physics of HFSWR dictates that many site-dependent factors contribute to the accuracy, reliability and availability of the various radar products. Moreover, the sensitivity to network configuration varies according to the type of measurement (or 'mission') being undertaken. Thus the choice of geographical sites which together comprise the network must reflect not only the family of radar outputs required but also their relative priorities. The resulting optimisation problem is extremely challenging.

An effective methodology for optimising HFSWR network design for the case where multiple missions must be addressed has been developed recently and demonstrated in the context of a hypothetical two-radar system deployed in the Strait of Malacca (Anderson, 2013). The results of that study demonstrated that quite disparate criteria can be accommodated within a genetic algorithm framework and confirmed that the method yielded the true optimum site configurations. Yet that study left a key question unanswered. In practice we are unlikely to be satisfied knowing that our choice of sites is the best for a given budget; we want to know that the network will meet prescribed levels of performance. This could well mean that, in a particular situation, a mix of quite different radar types would be required, adding another dimension to the network design problem.

In this chapter we review the genetic algorithm methodology for multi-objective optimisation in the HFSWR context, and show how it can be extended to handle the inverse problem of designing networks to meet specified performance levels. In order to illustrate the steps involved in formulating and applying the methodology, the discussion is framed in the context of a specific scenario: the design of an HFSWR network for providing surveillance and remote sensing of the South China Sea.

The spatial resolution and ultimate sensitivity of HFSWR is primarily a function of radar design, but performance in its various candidate roles is also dependent on a wide variety of geophysical factors, lithospheric, oceanic, atmospheric and ionospheric. Further, the relative priority of different missions reflects economic, geopolitical and strategic considerations. As all these aspects would (or should) be taken into account by network designers, it is appropriate to examine ways in which they can be incorporated in the objective functions employed for optimisation. In the following section we set the scene for the subsequent analysis by reviewing the physical environment and the associated human activities which an HFSWR radar network might be expected to monitor. Next we outline the capabilities and limitations of HFSWR in this context, based on the nominal performance of four existing radar systems. Once the radar capabilities have been established, we turn to the central issue, namely, that of formulating the network design problem in mathematical terms, which leads us to focus on evolutionary algorithms of nonlinear optimisation. Here the genetic algorithm approach of Anderson (2013) is emphasised, as it lends itself naturally to multi-objective optimisation, though in order to handle the enormous computational burden in the present case, a recently-reported convergence acceleration technique (Anderson et al, 2013) is introduced. We proceed to describe practical methods for constructing chromosomes and objective functions for a number of missions, illustrating these by relating them to the South China Sea context.

THE SOUTH CHINA SEA

Physical Geography

Formally the South China Sea extends from Bangka Belitung, between Sumatera and Borneo, to the northern extremity of Taiwan, and from the Gulf of Thailand to the Philippines, as shown in Figure 1.

Within its area of some 3,500,000 square kilometres lie several hundred islands, of which most are grouped into two clusters, the Paracel and Spratly Island chains. A great many of the islands are little more than exposed reefs and even the important Spratly Island group has a total land area of less than 5 square kilometres and a maximum elevation above sea level of only 4 metres. Some important features are entirely submerged, as is the case with Macclesfield Bank – actually an atoll-which is on average about 10 m below sea level, yet has an area of some 6500 square kilometres. Scarborough Shoal (akaPanatag Shoal), has reefs and small islets above water amounting to only a few hectares, surmounting an area of some 150 square kilometres of about 15 metres depth. beyond which the sea floor drops away rapidly to a depth of several kilometres. Only a handful of islands are large enough to be home to an airstrip and some of these facilities may not survive even a modest rise in sea level.

The large-scale bathymetry of the South China Sea is particularly striking. South of a line joining Brunei to the southern tip of Vietnam, the depth is less than 100 metres, but north of that line the sea floor descends rapidly to 1000 – 5000 metres, except around the island chains and atolls. With the exception of a few narrow but deep channels between Luzon and Taiwan, connecting to the East China Sea, the South China Sea is essentially a basin.

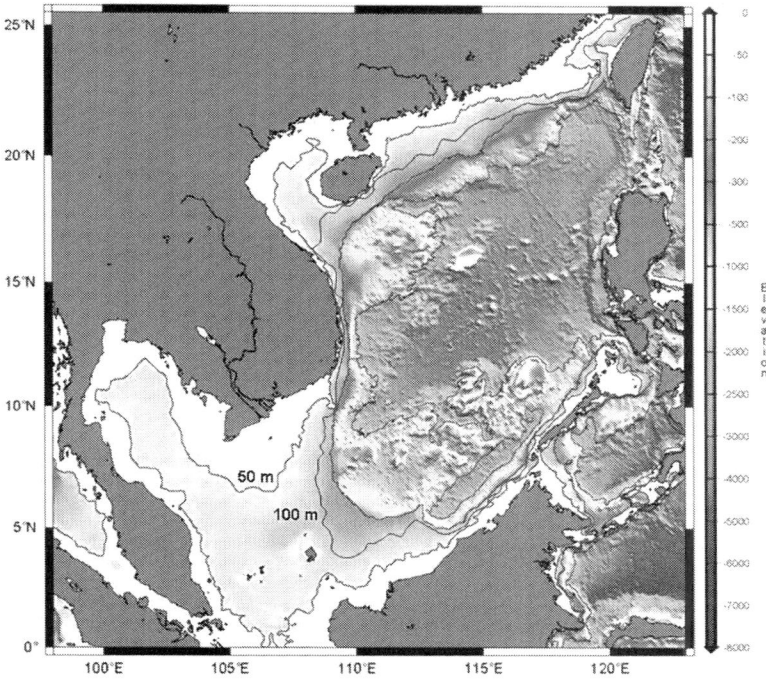

Figure 1: The bathymetry of the South China Sea (adapted from the World Data System for Marine Environmental Sciences, http://www.wdc-mare.org/).

Ocean surface conditions are influenced by the orography of adjacent land masses which helps steer the prevailing winds. In the case of the South China Sea the principal land feature that is relevant to HF radar system performance is the mountain range along almost the entire coast of Vietnam.

Meteorology and Oceanography

The wind regime over the South China Sea is dominated by the monsoon winds, punctuated by mesoscale systems such as tropical cyclones. During the boreal winter, the northeasterly winter monsoon winds impose a fairly uniform stress over most of the South China Sea, whereas in summer, June, July and August, the southwesterly monsoon winds show somewhat greater spatial variability, especially south of about 6°N. Average wind speeds in winter tend to fall in the range

8 – 12 m/s whereas the southwesterly summer monsoon winds are typically approximate 6 – 8 m/s in the Southern SCS and somewhat less in the northern SCS. Highly variable winds and surface currents are observed during the transitional periods. Moreover, synoptic systems often pass by the SCS and causes temporally and spatially varying wind fields. Severe weather most frequently takes the form of an increase in the strength of the prevailing monsoon winds or as meso-scale disturbances concentrated in either of two regions: a localised area east of the southern part of Vietnam, centred on 10°N, 110°N, and the band between Luzon and southern China. The mean wind regimes for summer and winter are shown in Figure 2.

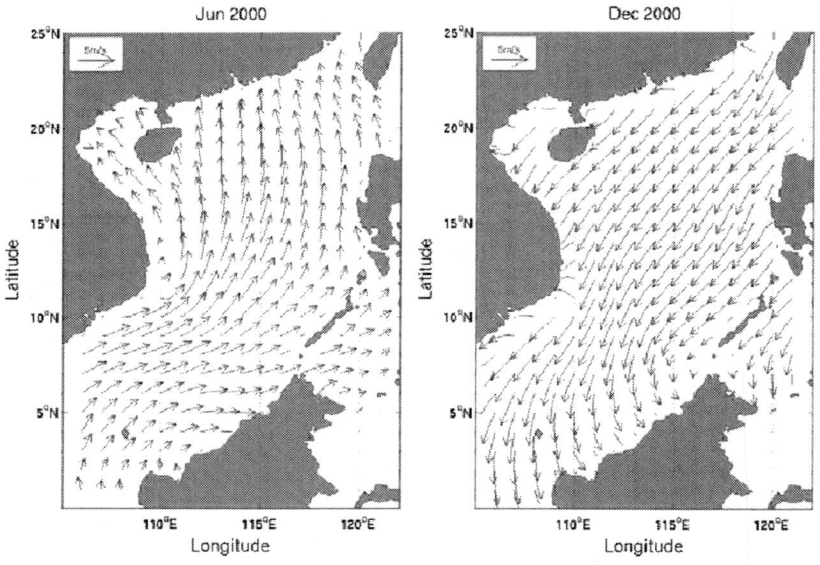

Figure 2: Synoptic-scale wind patterns during the summer and winter monsoon seasons (from Chu et al, 2003).

Tropical cyclones form in the waters between 12°N and 24°N, usually making landfall over Hong Kong and southern China, the north and central coasts of Vietnam or the northern Philippines. The most severe cyclones occur to the east of the Philippines and Taiwan, as shown in Figure 3 but, even so, the South China Sea north of 15°N.is occasionally subjected to category 3 and 4 events.

Wave and current distributions due to the wind forcing are less uniform than the wind fields. Significant wave height (SWH) distributions are higher in the northern and central SCS (north of 10°N) than in the southern SCS (south of 10°N) with upper quartile values exceeding 2.25 m. (The Wave watch III model has been found to yield fairly accurate results (Chu et al, 2004), so serves as a useful adjunct in modelling radar performance.) As shown in Figure 4, the orientation of the high SWH region coincides with the orientation of the monsoon winds.

Tropical Cyclones, 1945–2006

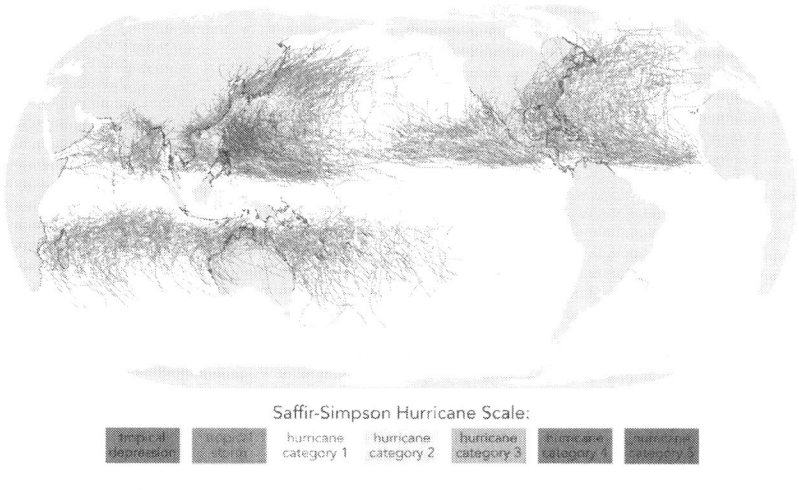

Saffir-Simpson Hurricane Scale:

Figure 3: The distribution of tropical cyclones over the period 1945 – 2006.

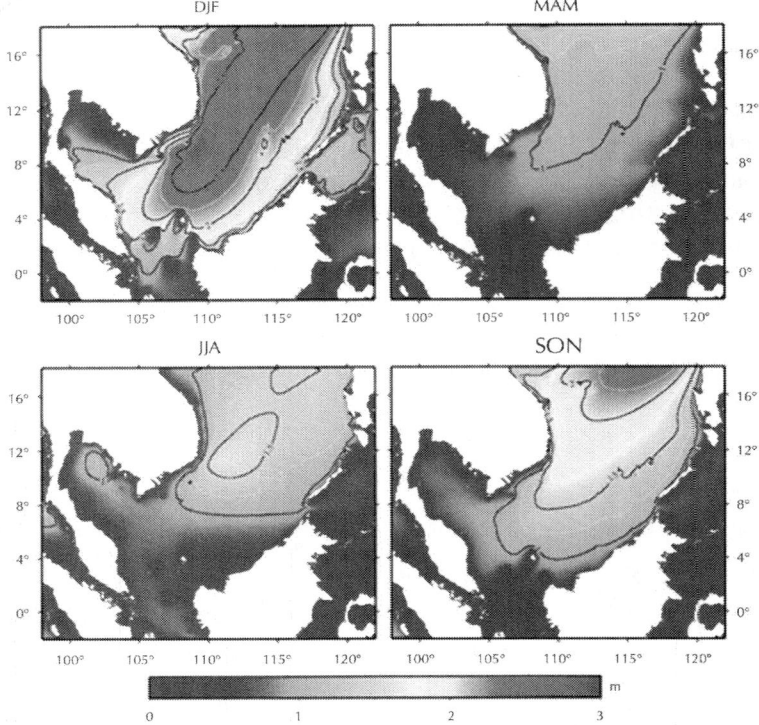

Figure 4: Seasonal variation of mean wave height for the period 1979-2009, WaveWatch III hindcast (Mirzaei et al, 2013). Note that the dominant wave direction is aligned with the monsoon winds, which is to say southward-propagating in DJF and northward-propagating in JJA.

The prevailing winds have a direct effect on the surface water currents of the shelf region. The current speeds are about 0.6 knots to the SW during the winter monsoon. They change to 0.2 to 0.4 knots to the NE during the summer monsoon. Stronger currents flow adjacent to the Vietnamese coast in particular, attaining speeds in excess of 1 m/s, while the islands of the Spratly archipelago can induce fairly complex local current variations. Primary or climatological current patterns in summer and winter are shown in Figure 5, but these convey an incomplete picture of the flow field. To gain a better understanding of the complexity of the current distribution, consider Figure 6, which shows the outputs of a detailed hydrodynamic model of the current field in the northern and central South China Sea. A key feature of this model is the inclusion of the wind-induced current, which was found to dominate the geostrophic current in many places. Moreover, note the appearance of the mesoscale eddies. These have been validated by observation. In a similar vein, Marghany (2009, 2011, 2012) has shown how local current patterns can be extracted from spaceborne SAR using observations off the east coast of peninsular Malaysia. The lesson to be drawn from this kind of modelling is that the flow field has significant structure on length scales of 50 km or less; given that the cross-range dimension of a typical HF radar resolution cell at long range may approach this magnitude, it is evident that HFSWR could provide unique validation data, though conventionally measured Doppler spectra will not always have discrete Doppler shifts and hence current velocity estimation will be compromised on those occasions.

One particular form of current perturbation which has received a lot of attention by the HF radar community is that associated with a tsunami (Lipa et al, 2012). It has been demonstrated that HFSWR is an effective tool for early warning of tsunamis provided that the bathymetry is favourable, which is to say reasonably shallow so that the speed of the tsunami is much reduced from its high deep water value. As Figure 1 shows, the southern part of the sea occupied by the Sunda Shelf and the north-western margins certainly satisfy this requirement.

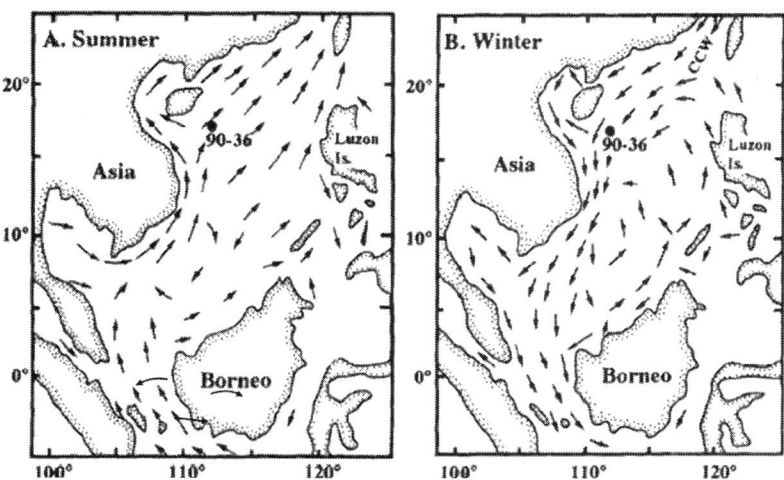

Figure 5: Primary currents during the summer and winter monsoon periods (Chen et al, 1985).

The SW winds blowing along the SW-to-NE part of continental shelf may induce upwelling during the summer, bring nutrients to the eutrophic zone on the outer portion of the shelf and, enhance primary production of the waters (Wang and Kester, 1988). The seasonal stratification stimulates the seasonal changes in primary production and nutrient cycling, with a strong signature evident in the high chlorophyll distributions in two coastal upwelling regions: the northwestern Luzon in winter and the eastern coast of Vietnam in summer. Mesoscale eddies provide another mechanism responsible for seasonal and interannual variability of the surface chlorophyll distribution.

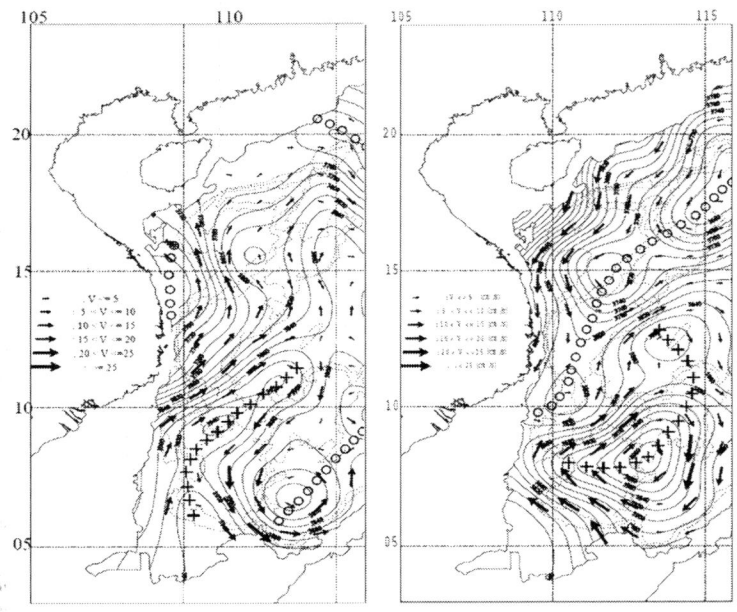

Figure 6: Modelled surface current fields for the central and northern South China Sea as computed with a 2-D numerical code (Ninh et al, 2000); (a) summer, and (b) winter.

The South China Sea surface layer is 50-100m thick. Its distributions are different in winter and summer. The monsoon-driven reversal of surface currents affects the temperature and salinity of the water masses, and hence the conductivity which impacts on HF radar performance. In winter, due to the influence of the northeast monsoon, the temperature increases progressively from the coast to the outer sea and the salinity decreases progressively from north to south. The temperature ranges from 22°C in the north to 26°C in the south, while the salinity varies from 33.2 to 34.5 PSU. In summer, due to the influence of southwest monsoon, the surface temperature is generally 28-29°C and the salinity is low – near 32 PSU – in the north and south, and high – about 33.6 PSU – in the central region. In summer, the SW monsoon results in the large increase in rainfall and river discharges. This results in the reduction of salinity in the coastal waters and the production of seasonal pycnoclines. Particularly low salinity occurs off the east coast of peninsular Malaysia.

The diurnal and semi-diurnal tides are of about equal magnitude in the South China Sea, though the latter is more effective at generating internal waves. These are exceptionally strong in the northern region, where they are generated in the Luzon Strait before propagating westward. Amplitudes reach 200 m with horizontal scales upwards of 200 km. These internal waves take about 4 days to cross the South China Sea, modulating the surface gravity wave field as they progress and hence influencing the radar scattering properties of the sea surface.

Shipping

The volume of shipping activity in the South China Sea can be illustrated by a few key statistics:

- nearly half the world's annual merchant fleet tonnage moves through its waters, carrying commodities valued at over $5 trillion
- one third of global oil tanker traffic and over half of global LNG traffic crosses the South China Sea, most from the Strait of Malacca but the very largest supertankers via the Sulu Sea
- ore carriers, predominantly from Australia, transport roughly half a billion tonnes of iron ore and a similar amount of coal through the South China Sea annually
- six of the world's ten largest ports lie on the coastlines of the South China Sea
- the annual growth rate for liquid petroleum fuels consumption in recipient countries – mainly China and Japan and South Korea – is presently 2.6 %, while that for natural gas is 3.9 %. For Australian minerals the figure is 4.6%

- over half a billion people live within 100 miles of its margins
- perhaps as many as 18,000 small fishing boats ply its waters

The major shipping routes are shown on Figure 7, using data derived from Wang et al, 2013).

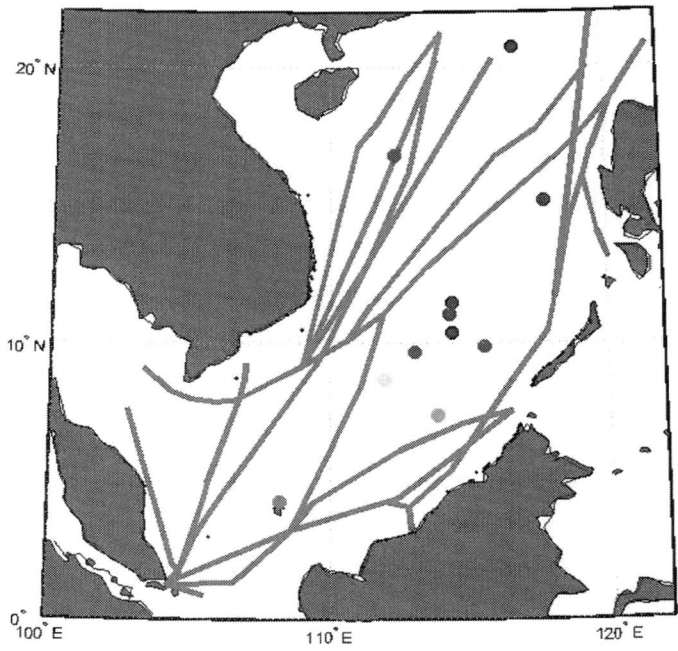

Figure 7: Principal shipping lanes through the South China Sea, and islands selected for use in radar mission definitions.

Given the density of traffic, it is perhaps not surprising that shipping hazards in the South China Sea continue to take a toll on vessels in transit, as exemplified by several recent incidents: the sinking of the Bright Ruby (severe storm, November 2011), Royal Prime (hit reef, and sank, December 2012), Harita Bauxite (sank after engine failure, February 2013), Jung Soon (sank after hull failure, September 2013). Another form of hazard is piracy, for which the South China Sea was once notorious. While less frequent than a few years ago, hijacking and armed robbery remain a significant threat in some waters. Mimicking the 'mothership' refuelling station tactic used by pirates off the coast of Somalia, pirates in Indonesia and Malaysia tend to camp on a small island near to narrow shipping lanes and launch their strikes from there. Pirates in South East Asia also tend to launch their attacks at night, which makes it much harder for ship captains to spot them coming. Between 2008 and 2010, 57 incidents of 'cluster piracy' took place around the Abambas / Natuna/ Tambalan corridor. In the first six months of 2013, attacks involving pirates boarding vessels and assaulting the crew were recorded in the Singapore Straits, in Malaysian waters, in the Straits of Malacca and in the Philippines. Within the main body of the South China Sea, 2013 escaped serious incident.

Another consequence of heavy ship traffic is oil pollution, both accidental and deliberate, such as that caused by tankers flushing their tanks on the voyage back to the Middle East. Offshore facilities and undersea pipelines are other man-made sources.

Economic Activity

It is evident from the preceding discussion that 'through traffic' is critically important to the destination countries of China, Japan and South Korea, but, for the littoral states around the South China Sea, fishing is the most vital maritime activity, as it has been for centuries. Fish protein constitutes nearly a quarter of the average Asian diet and demand continues to grow strongly. Thus, whereas the extent of oil and gas reserves beneath the South China Sea remains questionable, the value and importance of its fisheries and aquaculture is not in doubt. It is therefore of great concern that the relative stability of traditional fishing practices is now threatened by over-fishing, together with rising water temperatures which appear to be resulting in migration of fish populations, primarily further north. These developments are stoking tensions between the countries whose populations depend on accessible and reliable stocks.

It is widely reported that the South China Sea holds immense untapped natural reserves of oil and gas, and that the contested ownership of the Spratly Islands and other parts of the sea is primarily a fight for these resources. It is certainly the case that confrontation and armed skirmishes have taken place where exploration has been pursued in disputed waters. Yet a considered analysis does not support the more extreme assertions regarding the magnitude of the reserves in the contested regions. The most recent assessment by the US Energy Information Administration estimates that the total of proven and probable reserves in South China Sea amounts to approximately only 11 billion barrels of oil and 190 trillion cubic feet of natural gas. Another US expert source places the figure for oil at 2.5 billion barrels. Allowing for additional reservoirs in under-explored areas, the EIA says, could add between 5 and 22 billion barrels of oil and 70 to 290 trillion cubic feet of gas. These figures contrast with those of the Chinese National Offshore Oil Company which estimates undiscovered reserves amount to 125 billion barrels of oil and 500 trillion cubic feet of natural gas. In the absence of detailed prospecting, the actual quantities cannot be known with any certainty. What is undeniable is that the preponderance of known resources resides in the uncontested areas close to the coasts of the surrounding countries, especially Vietnam, Malaysia and Brunei. Thus the fierce competition for control, if not ownership, of the islands, reefs and shoals of the South China Sea, is probably driven by a combination of factors, economic, political and strategic.

We note too that Malaysian researchers have identified potentially valuable elements in seabed sediment, including manganese, zinc, chromium, lead, copper and aluminium.

Strategic and Geopolitical Issues

The South China Sea has a long history of tension and conflict. Much of this derives from overlapping territorial claims and disputed ownership of maritime features, as indicated in Figures 7 and 8, exacerbated by a race to exploit the maritime zone's natural resources. In addition, there is also a growing element of overt strategic rivalry and nationalism, which poses a substantial risk to regional security and prosperity. Specific areas of dispute include:

- the Spratly Islands, disputed between the People's Republic of China, the Republic of China, and Vietnam, with Malaysia, Brunei, and the Philippines claiming part of the archipelago
- the Paracel Islands, disputed between the People's Republic of China, the Republic of China, and Vietnam
- the Pratas Islands, disputed between the People's Republic of China and the Republic of China

- the Macclesfield Bank, disputed between the People's Republic of China, the Republic of China, the Philippines, and Vietnam
- the Scarborough Shoal, disputed between the People's Republic of China, the Philippines, and the Republic of China.

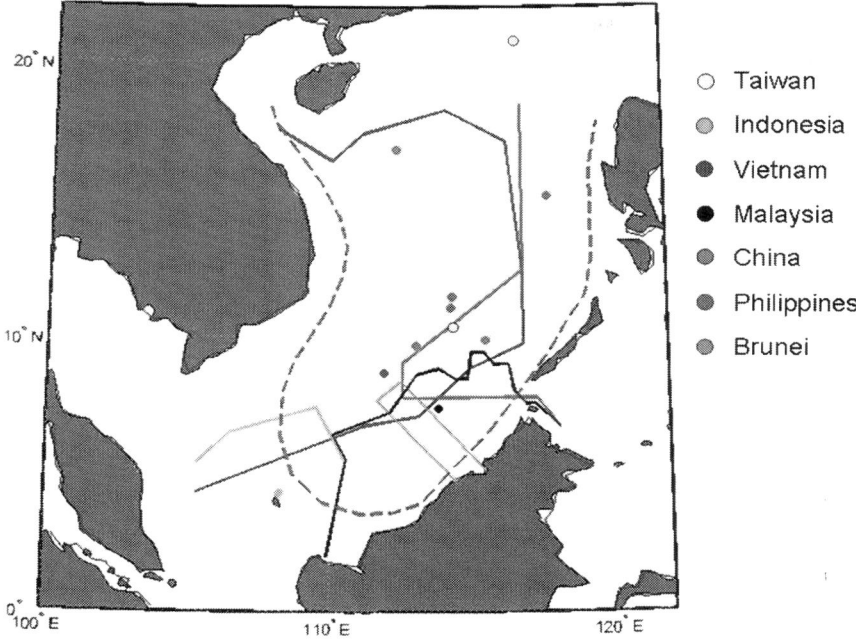

Figure 8: Territorial claims over the South China Sea, together with occupied islands.

Many detailed discussions of these issues can be found in the open literature (International Crisis Group 2012a, 2012b); here it suffices to make the point that timely, comprehensive, robust and persistent surveillance can be a useful means of establishing trust and defusing incidents which could spiral out of control.

CAPABILITIES AND LIMITATIONS OF HF SURFACE WAVE RADAR

The Suitability of HFSWR for Maritime Remote Sensing and Surveillance

The physical quantities which impact directly on HF radar capability in both its remote sensing and surveillance roles are (i) water electrical conductivity, (ii) surface currents, and (iii) the geometry and dynamics of the sea surface, usually represented as a spectrum of surface gravity waves. It is no exaggeration to state that radar performance in any of its missions is highly dependent on these primary quantities. Moreover, the primary quantities are coupled with other geophysical variables and processes, as illustrated in Figure 10. Therefore, as part of the network design procedure, it is absolutely essential to take into account the kind of information presented in Section 2. Figure 10shows the linkages between the primary quantities and other geophysical variables and processes. From this figure it is apparent that, by appropriate analysis and interpretation of the radar echoes, it may be possible in some circumstances to use the primary measurements of currents and wave spectra to infer secondary phenomena, surface winds being the prime example.

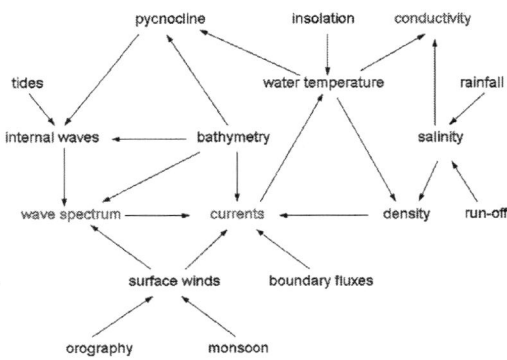

Figure 9: The relationships between the geophysical parameters and phenomena which impact on HF radar performance.

It is evident from the discussion in Section 2 that real-time monitoring of these environmental conditions and ship traffic over the South China Sea could have substantial value for a wide range of users. Yet the practical application of any remote sensing technology requires that we first establish whether the coverage, resolution and accuracy of the measurements are commensurate with the needs of the users. To illustrate this step, we shall consider the nominal performance of four well-known HFSWR products.

The commercial marketplace for so-called 'oceanographic' HFSWR systems is dominated by two manufacturers: CODAR, with its Seasonde radars (Barrick, 1998), and Helzel Messtechnik, with its WERA systems (Helzel et al, 2010). These radars each cost in the vicinity of 0.5M$ per system consisting of one transmitting station and one receiving station, and have excellent track records for delivering ocean current information at ranges out to 200 km, with sea state measurements available for significantly shorter ranges. While some extravagant claims are made about the ability of these low power radars to detect ship targets at ranges of several hundreds of kilometres, experience has tended to show that reliable detection is confined to 50 – 150 km, depending on target type, time of day, and other factors. These radars have quite different characteristics so it is appropriate to include them both in the list of options for a heterogeneous network.

The South China Sea is roughly 1000 km east-to-west at 10° N so it is clear that oceanographic radars based on undisputed territories are not able to provide comprehensive surveillance. The possibility of deploying radar systems on small islands may change this assessment somewhat, as discussed later in this paper, but a priori it would seems that radars with far superior long range performance are required if comprehensive surveillance is to be achieved. For the study reported here, two commercial HFSWR radars were chosen to represent such 'military-class' systems: Raytheon's SWR-503 (Ponsford, 2012) and Daronmont Technology's SECAR radar (Anderson et al, 2003). Each of these systems has demonstrated ship detection at ranges well in excess of 400 km. It is important to point out that the diverse observations and opinions from which these figures were inferred correspond to a variety of environmental conditions, so the estimates are really just indicative. Still, the table serves to provide numerical values for the purpose of exercising the network optimisation suite.

Table 1: Capabilities of selected HFSWR systems, expressed in terms of typical maximum ranges at which measurements can be made reliably

Observable	Typical performance			
	low-cost civilian radar		military radar	
	Max. range (km)	accuracy	Max. range (km)	accuracy
surface current	60 - 200	± 0.02 – 0.20 m/s	350 - 450	± 0.02 – 0.10 m/s
wave height	30 - 100	± 10 – 25 %	150 - 350	± 10 – 20 %
wind direction	50 - 180	± 30° - 60°	320 - 400	± 20° - 30°
wind speed	30 - 150	± 20 %	150 - 350	± 20 %
large ship	50 - 180	± 0.5 - 3 km	300 - 450	± 0.5 – 3 km
fishing boat	20 - 65	± 0.5 – 2 km	120 - 280	± 0.5 – 2 km
small boat	10 - 45	± 0.5 – 1 km	70 - 150	± 0.5 – 1 km

A natural first test is to see whether even the most potent (and expensive) network could deliver the desired coverage. In order to obtain a large and realistic set of possible sites in the present case, we visually searched the coastlines around the South China Sea as presented in Google Earth™, selecting as candidate locations all those places characterised by reasonably flat, low-lying ground with linear sea frontage in excess of 300 m. These criteria were applied to ensure a choice of radar type at every location; only the CODAR Seasonde is able to be deployed on almost any topography. Some 141 sites emerged from this procedure; they are marked on Figure 10.

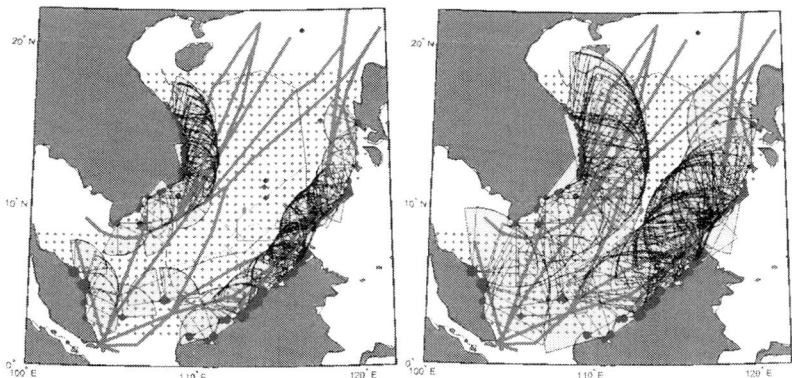

Figure 10: Maps of the South China Sea showing the major shipping lanes, candidate radar sites (blue dots), the associated potential radar coverage (yellow sectors), the discrete points in the sea area at which objective functions can be evaluated (magenta dots), and selected islands of interest (dots, various colours). Figure 10a shows radar coverage to 200 km, Figure 10b to 400 km.

Figure 10a shows the nominal current measurement coverage for oceanographic-class radars deployed at each of these sites, while Figure 10b shows the corresponding information for a military-class radar. These figures reveal that no fully-compliant solution exists, even with the maximal deployment of radars, but they suggest that a solution employing a combination of radar types, at a suitable subset of sites, might achieve an acceptable outcome leaving relatively few areas unsurveyed for this particular mission.

Regarding spatial resolution, while the cross-range dimension of a cell at nominal maximum range exceeds the along-range dimension by up to an order of magnitude, the broad features of oceanographic fields remain distinguishable, and discrete ship echoes can be finely resolved in the Doppler domain, so HFSWR is certainly able to provide the required detail for most objectives. The accuracy of measurements is limited not so much by radar design as by the intrinsic spatial and temporal variability of natural phenomena; the widespread acceptance of HFSWR remote sensing products confirms that the information is of adequate fidelity.

Performance Limitations and Constraints

It is helpful to be aware of the factors which limit HFSWR performance, as radar and network design can be adapted to minimise the deleterious effects of some of them. First there is the nature of surface wave propagation, which results in increasingly rapid signal decay as one moves beyond about 50 km range, with higher frequencies decaying much more quickly. Second, there is the frequency dependence of the radar signatures of both ship targets and the ocean surface, which have complicated forms that jointly have a strong influence on radar performance. Third, there is the external HF noise from lightning and man-made emissions, which almost always defines the noise floor against which radar echoes of interest must compete. External noise is highly dependent on time of day. Fourth, there are Doppler-spread echoes from the ionosphere, which have a complex spatial and temporal pattern of occurrence and can mask echoes of interest.

Measures which can be taken to mitigate these factors include antenna design, advanced signal processing, frequency agility and, of special relevance to the present study, siting relative to the locations and velocities of the phenomena under observation. As a simple example, Figure 11 shows the obscuration of ship echoes due to sea clutter, plotted in Doppler space.

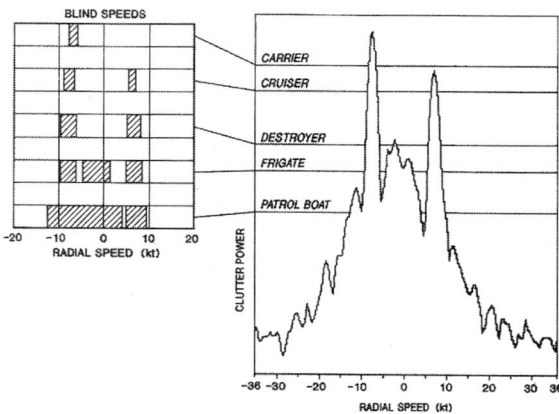

Figure 11: Blind speeds for various ship types, against a specific sea clutter spectrum, for a 32 second integration time.

RADAR SITING AND CONFIGURATION DESIGN AS A MULTI-OBJECTIVE OPTIMISATION PROBLEM

Elements of the Formulation

At the outset we need to identify the data structures, procedures and supporting information that need to be integrated into the problem formulation. These are:

- the parameter space $\boldsymbol{P}$ in which the solutions must lie. A particular solution $x_{mn} \in X_{mn}$

  will have a fixed number m of transmitting systems, each specified by location, orientation and design, together with a fixed number n of receiving systems, each similarly described by its location, orientation and design. The ability of receiving system p to acquire and process signals from transmitting system q is represented by a coupling matrix C^{mn}:

$$C^{mn}_{pq} = \begin{cases} 1 \text{ if receiver p can process signals from transmitter q} \\ 0 \text{ otherwise} \end{cases}$$

Usually compatibility demands that the systems belong to the same product family. Thus, for example, a Seasonde receiver can process signals from a Seasonde transmitter, but not signals from a WERA transmitter. In the problem under consideration, we do not know a priori what the network membership numbers m and n should be; accordingly we define $\boldsymbol{P}$ as the disjoint union of the X^{mn},

$$P = \cup_{n=1}^{N} \cup_{m=1}^{M} X^{mn}$$

Where M and N are upper bounds on the numbers of radar transmit and receive sites.

A solution x thus can be written in the form of a pair of two-dimensional arrays,

$$x \triangleq \left[\left\{ R_j^{lat}, R_j^{lon}, \Psi_j, \varphi_j \right\}_{j=1,N} ; \left\{ T_k^{lat}, T_k^{lon}, \Psi_k, \varphi_k \right\}_{k=1,M} \right]$$

Where the dimensions correspond to parameter type (latitude, longitude, radar class, orientation) and parameter index (labelling the set of Tx/Rx sites which make up the configuration)

- The specific coordinates of feasible sites. These constitute a subset of the set of points C which comprise the coastlines which border the South China Sea or, in the case of radar sites on very small islands, the nominal location at which the installation is most feasible

- the amenity of each location to the installation of a radar, taking into account factors such as accessibility, power supply, field of view and environmental impact

- the range and azimuthal coverage of the individual radars to be used

- the wind, wave and current climatology of the waters of the South China Sea

- the recognised shipping lanes

- the types of vessels of interest and their typical speeds and radar cross sections

- the surveillance and remote sensing missions assigned to the radar system and the associated performance thresholds which must be exceeded (at least in a statistical sense)

- algorithms which compute the radar network response for any given combination of ship type, course, speed and environmental conditions

- the objective function space Y, that is, the k-dimensional space whose coordinates measure the radar performance against the k tasks assigned to the radar

- a search algorithm which finds the extrema of a scalar function over a specified domain
- a criterion for ranking solutions which achieve extrema in one or more coordinates of Y

It is common practice to formulate optimisation problems in terms of minimising the objective functions rather than maximising them, which is trivially achieved by redefining the coordinates of Y; we shall follow this practice.

Criteria for Network Optimality

With this palette of ingredients, various radar siting problems with cost and performance constraints can be formulated. Three of the most important are:

- find the solution $\bar{x} = x^{mn} \in X$ which maximises performance against a specific task for a given cost
- find the minimum cost solution $\bar{x} = x^{mn} \in X$ which exceeds a specified threshold of performance
- Given an existing deployment x_{old} of a number of radars illuminating parts of an area of interest, find the solution xn_{ew} such that the augmented network $\bar{x}_{aug} = x_{old} \cup x_{new}$ meets some specified performance/cost criterion but there are many other possibilities. We observe that some of these can be expressed as inverse problems with the threshold vector taking the role of the data vector.

Multi-objective Optimisation via Pareto Dominance

The definition of the problem given above is in one sense incomplete – it does not specify the choice of norm for the space Y. In a single objective optimisation problem, the objective space is usually a subset of the real numbers and a solution $x_1 \in P$ is better than another solution $x_2 \in P$ if y1<y2

where $y1=\mu(x_1)$ and $y2=\mu(x_2)$. In the case of a vector-valued objective function mapping,

comparing solutions is more complex and one must endeavour to capture the essential priorities of the problem in the choice of norm. Herein lies the crucial distinction between single objective and multi-objective problems-whereas the former afford simple scalar measures of fitness that can be used to rank individual members of the design space, the latter are characterised by conflicts of interest among the competing objectives as measured by $\mu_i, i=1, m$.

There are several ways to deal with this complication. Perhaps the simplest is to create a scalar figure of merit as a weighted sum of the separate objective measures,

- minimize $\mu^{(1)} = \sum_{i=1}^{m} \alpha_i \mu_i$

Another approach is to convert all but one of the objectives into constraints,

- minimise μj subject to $\mu_i \leq z_i \forall i = 1, m; i \neq j$

While convenient, these methods shed little light on the nature of the trade-offs made. As there may be subtle, non-quantifiable considerations involved in site selection, such as risks to personnel or to equipment, a better approach is to map the trade-off surface so that the decision maker can execute judgment in making a final selection. To perform this mapping, it is not necessary to run (i) or (ii) above for a large number of parameter selections α_i, z_i and to inspect the outcomes. Instead, we can use an evolutionary stochastic optimisation algorithm to reveal the Pareto front, as described below.

Pareto optimality is based on the binary relation of dominance. A solution $x_1 \in X$ is said to be dominated by another solution $x_2 \in X$, written $x_2 < x_1$, if x_2 is at least as good on all counts (objectives) and better on at least one, that is, $\mu_i(x_2) \leq \mu_i(x_1) \forall i=1, m$ and $\mu_j(x_2) < \mu_j(x_1)$ for some j. With this relation, the Pareto set of optimal (non-dominated) solutions P^* will usually have multiple entries, associated with different trade-offs between the objectives. The image $Y^* \subset Y$ of the Pareto set $P^* \subset P$ is referred to as the Pareto front and knowledge of its shape greatly assists in choosing the best compromise solution.

Implementation via Genetic Algorithms

Classical techniques for finding extrema of functions defined on prescribed domains rely, in most cases, on gradient search methodologies. Such techniques are vulnerable to being trapped on local extrema, rather than the global extremum of main interest. In addition, the convergence may be slow, especially near the extrema, necessitating the invocation of higher-order derivatives. While there are ways to alleviate these weaknesses, they come at considerable cost. An alternative approach, now in widespread use, is to emulate evolutionary mechanisms which we observe in action in the natural world. The best known of these evolutionary optimisation techniques are genetic algorithms.

Genetic algorithms encode the parameter values associated with each candidate solution as a string, usually in binary format. For each parameter, the number of bits provided must be sufficient to encode the full range of possible values associated with that parameter. The string representing a solution is simply the concatenation of the sub-strings corresponding to the individual parameters; by analogy with biology, this string is referred to as a chromosome. Starting with an initial population of candidate solutions (ie, chromosomes) constructed by means of a random number generator, a genetic algorithm iteratively applies three basic steps: (i) rank the members of the current population according to fitness, (ii) select superior members which will be used to breed the next generation, and (iii) apply operators on randomly-selected pairs of these members to mimic the transfer of genetic material to offspring that occurs during biological reproduction, thereby producing a new generation with statistically superior characteristics.

A common mechanism for the transfer of information from one generation to the next is variable length cross-over. For each pair of chromosomes selected to breed together, the start and end indices of a sub-string are selected by a random number generator and the corresponding sub-strings are exchanged. The excisions are not forced to align with the parameter sub-string boundaries. The offspring of this coupling have parts in common with each parent, and in general will represent new solutions. A small fraction of this new set of chromosomes is then subjected to mutation, that is, one or two bits may be flipped to produce a different string, which of course maps onto a different candidate solution. This completes the process of constructing a new generation.

With single objective optimisation, it is a simple matter to rank the members of the resulting population so that selection of candidates for constructing the next generation can proceed. Chromosomes representing the best solutions are carried over unchanged to the next generation, as well as participating in the breeding cycle, while the least fit are discarded. The resulting population is then allowed to breed in its turn, via cross-over and mutation. After passing through a large number of generations, the population tends to converge towards a uniform composition whose members share the most desirable parameter values. Importantly, by virtue of the randomness of the cross-over and mutation operations, candidate solutions from all over the solution domain are potentially represented, and mutation ensures that this property is maintained, so that the population is unlikely to be trapped on a local extremum if a superior solution exists.

With multi-objective optimisation, the key objective is to find the Pareto front, but experience has shown that coverage and convergence can be improved by relying on more than just Pareto dominance for selection. In our approach, each chromosome was tested against its contemporaries and those which were Pareto dominant were automatically selected, while those which had only one or two dominators were also short-listed. In addition, members that performed particularly well against just one objective function were retained. Supplementing these criteria, a scalar figure of merit was defined by taking the product of the individual objective functions; this provided another metric for selection. The total size of the population was maintained at the initial value by allowing each of these different selection mechanisms to contribute a fraction of the membership, with the relative proportions changing with the generation index. We modified the single objective genetic algorithm developed by Anderson (2013) to embody these ideas and hence to compute an estimate of the Pareto front.

Acceleration Techniques

Genetic algorithms tend to be computationally expensive, so special techniques continue to be developed to accelerate convergence. Some methods which have proven efficacious are:

- eugenics – a recent hybrid scheme which combines the virtues of GA with a very efficient gradient search
- class identifiers – partitioning chromosome space into dissimilar clusters and constraining cross-over to avoid in-breeding, thereby increasing and maintaining diversity
- smart seeds – using intuition, experience and common sense to insert some chromosomes with high potential

Methods for Handling Variable Solution Space Dimensionality

One of the challenges of the general network optimisation problem is that, unlike the case in Anderson (2013), the number of radars is itself a variable. Given that the gene length for representing an individual radar is fixed, it follows that the minimum chromosome length will change. This introduces some very fundamental modifications to the elements of the GA algebra, so a number of approaches have been explored:

- loop through dimension index; a straight-forward extension of conventional GA structure
- set the chromosome length to the maximum number of radar sites considered feasible and work within this space; likely to be computationally expensive
- adopt a hierarchical scheme, with fixed length chromosomes containing genes serving as pointers to subspaces of different dimensionality; potentially effective but complicated
- employ variable length chromosomes; this requires a whole new class of genetic operators able to work with strings of different lengths

In the implementation we have used for the South China Sea example, the first of these options has been adopted.

Constructing the Chromosomes

The chromosomes do not need to encode all the detailed information about site properties, radar characteristics, and so on. It is more efficient to use the genes as pointers to data files in which the numerical specifications are stored. In our illustrative example, we allow for four different radar types, so 2 bits are required for that purpose. Our survey of the coastlines of the South China Sea identified 141 candidate sites, so it might seem logical to allocate $\log_2 141$ bits to represent them. This causes a problem, as not all 8-bit strings correspond to radar sites, and the extra algorithmic structure that would be required to deal with this issue would arise in a section of the code which is run intensively. In the present case it is better to prune the set back to 128 sites, with minimal impact on the outcome, though conceivably another problem could justify increasing the state space to 256 sites. Thus our basic gene has $2 + \log_2 128 \equiv 9$ bits. As it necessary to extract the separate radar-type and radar-site parameters, an efficient 'gene scissors' is required, easily implemented in Matlab.

The specific context imposes other constraints that need to be carefully considered. For instance, in the present network design study, we found candidate sites located on the mainlands of Vietnam, Malaysia, Brunei and the Philippines, as well as on several islands some of which are of disputed sovereignty. It may be that network operations embracing radars in all ASEAN nations could be negotiated, but such arrangements area never simple. The situation becomes even more complicated when we contemplate radars on those islands which presently are home to airstrips, ideal for basing array-type HFSWR systems, since islands meeting that description are owned or occupied by China, Indonesia, Malaysia, the Philippines, Taiwan and Vietnam. In addition, a number of mostly submerged reefs and seamounts in the Spratly Islands bear constructions on which CODAR Seasonde radars could easily be fitted. All these possibilities need to be taken into account when proposing the extent of the solution space in which the set of optimal solutions is to be sought.

CONSTRUCTING THE OBJECTIVE FUNCTION SPACE

Objective Functions for Priority Missions

While it is certainly possible to conceive of many useful missions which could be addressed by a network of HFSWR systems, it is generally the case that one focusses on those which have a high level of economic or geo-strategic relevance. As an example, the palette of tasks which one might wish to address could take the form:

- maintain surveillance around most, preferably all, of the important islands with the targets of interest being vessels of at least patrol boat size, typically 50 – 80 m in length and long endurance research ships of around 100-120 m length

- provide full ocean current vector information over those parts of the South China Sea which are traversed by large vessels such as tankers and container ships; with emphasis on the major shipping routes

- provide sea state information for the areas in which fishing fleets operate

It is readily seen that the spatial domains over which the performance of these three tasks is of interest are of different dimensionality. As the objective function used to define fitness for a given task involves integration over the corresponding domain, there is a strong link between task domain and computational load. The cases of most concern to the network optimisation problem under consideration are as follows:

Domain dimensionality	Examples
0	islands, shoals, offshore oil platforms,
1	shipping lanes, transects, sovereignty and EEZ boundaries
2	fishing grounds, oil exploration leases, wave energy surveys

For each of the designated tasks, it is necessary to define some criterion that quantifies performance and which can thus be used to govern the search for the Pareto optimum configurations. To illustrate, we shall outline the construction of objective functions – also known as fitness functions or figures of merit – for the first two tasks mentioned above.

Ship Detection

Suppose an HF surface wave radar operating at a fixed centre frequency *f* is deployed with the goal of detecting ships whose radar cross section (RCS) exceeds some specified threshold. For detection we require that a ship echo exceed the clutter and noise power in the same Doppler bin by some margin ε, that is, there exists $\omega \in [-\Omega, \Omega]$ such that $s(\omega) > c(\omega) + n(\omega) + \varepsilon$ where $s(\omega)$, $c(\omega)$ and $n(\omega)$ are the target, clutter and noise power spectral densities respectively and $[-\Omega, \Omega]$ is the extent of the Doppler domain. At modest ranges, dependent on the radar type, the clutter power spectral density exceeds that of external noise, but at longer ranges external noise dominates and sets the detection limit. This we need to have a database which provides these distributions. From the description in Section 2 we know that the wind stress and hence the sea state is relatively constant over the South China Sea during each of the two monsoon periods, comprising some 80% of the year, so a reasonable approach is to compute clutter Doppler spectra for just these two sea states. In the context of large ships on the major lanes in the South China Sea, proceeding along known shipping lanes at fairly uniform speeds, $v \in [v_{min}, v_{max}]$, the Doppler perceived by a radar from a given ship is a function of a single coordinate, representing the ship's position along its chosen lane, since that determines the viewing geometry. Accordingly, for these targets it makes good sense to define a figure of merit which measures the fraction of the time (equivalently distance along the route under surveillance) for which such ships are detectable. In the case of small or medium size ships near particular islands or facilities, the direction of travel and the speed cannot be assumed, so the figure of merit should reflect the need to maximise detectability against all eventualities. Assuming a maximum speed V_{max}

$$OF_1 = \frac{1}{2\omega_m} \sum_{k=1}^{n} \int_{-\omega_m}^{\omega_m} H\left(s(\omega; r_k) - c(\omega; r_k) - n(\omega) - e\right) d\omega$$

(1)

where H(x) is the Heaviside function,

$$H(x) = \begin{matrix} 0 & x < 0 \\ 1 & x \geq 0 \end{matrix}$$

And

$$\omega_m = \frac{2v_{max} f}{c}$$

Where f is the radar frequency and c the speed of light. The r_k are the coordinates of the discrete islands, offshore oil platforms or other discrete features of interest.

It is a computationally trivial but operationally useful generalisation to apply a priority weighting to the individual islands,

$$OF_2 = \frac{1}{2\omega_m} \left(\frac{1}{\sum_k w_k} \right) \sum_{k=1}^{n} w_k \int_{-\omega_m}^{\omega_m} H\left(s(\omega;r_k) - c(\omega;r_k) - n(\omega) - \varepsilon \right) d\omega$$

(2)

Which could reflect the distribution of navigation hazards, risk of piracy, cross-Strait traffic density and so on. To evaluate these integrals, we need expressions for $s(\omega;r)$ and $c(\omega;r)$, as well as noise data. The first of these can be written

$$s(\omega;r) = R(\psi_{Rx}) \left(\frac{c^2}{4\pi f^2} \right) G(r_{Rx}, r) \sigma(\omega;\varphi_{scat}, \varphi_{inc}, r) \delta(\omega - \omega_D) G(r, r_{Tx}) T(\psi_{Tx}) P_{Tx}$$

(3)

with P_{Tx} the transmitted power, $T(\psi_{Tx})$ and $R(\psi_{Rx})$ denoting the azimuthal gain patterns of the transmit and receive antennas, $G(r_2, r_1)$ representing the propagation loss factor between positions r_1 and r_2, and $\sigma(\varphi_{scat}, \varphi_{inc})$ the bistatic radar cross section for an incident angle φ_{inc} and scattered angle φscat as defined at r , and ω_D the Doppler shift associated with the target echo,

$$\omega_D = \frac{-f}{c} \times \frac{d}{dt}\left(\left|r - r_{Tx}\right| + \left|r - r_{Rx}\right|\right)$$

(4)

For target-specific criteria, the RCS must be calculated using a computational electromagnetics code such as NEC4 or FEKO™.

The corresponding expression for c(ω;r) takes the form

$$c(\omega;r) = R(\psi_{Rx})\left(\frac{c^2}{4\pi f^2}\right)G(r_{Rx},r)\sigma(\omega;\varphi_{scat},\varphi_{inc},r)G(r,r_{Tx})T(\psi_{Tx})P_{Tx}A$$

(5)

Here A denotes the area of the resolution cell, whose cross-range dimension increases with range from the receiver. The cell's range extent is determined, in general, by the bandwidth B of the transmitted waveform and, for a phased array system of aperture L_{Rx}

we can write

$$A \approx \frac{c^2\left|r - r_{Rx}\right|}{2BL_{Rx}F\cos\psi_{Rx}}$$

(6)

The sea surface scattering coefficient

$\sigma(\omega;\varphi_{scat},\varphi_{inc},r)$ has a continuum of spectral content and, being dependent on sea state, will normally vary with position.

$$\sigma\left(\omega;\varphi_{scat},\varphi_{inc},r\right)=2^6\pi\,k_0^4\left[\sum_{m=\pm1}\int S\left(m\kappa;r\right)\delta\left(\omega-m\sqrt{g\kappa}\right)\delta\left(\kappa+k_{inc}-k_{scat}\right)d\kappa\right.$$

$$+\sum_{m_1=\pm1}\sum_{m_2=\pm1}\iint\Gamma^2\left(m_1\kappa_1,m_2\kappa_2\right)S\left(m_1\kappa_1;r\right)S\left(m_2\kappa_2;r\right)$$

$$\left.\times\delta\left(m_1\kappa_1+m_2\kappa_2+k_{inc}-k_{scat}\right)\delta\left(\omega-m_1\sqrt{g\kappa_1}-m_2\sqrt{g\kappa_2}\right)d\kappa_1d\kappa_2\right]$$

(7)

where $S(\kappa;r)$ is the directional wave spectrum at location r and $\Gamma^2(m_1\kappa_1,m_2\kappa_2)$ is a kernel which contains, inter alia, the polarisation dependence, though that does not play a role here. An segment of the resulting database of Doppler spectra, for a given frequency, evaluated for all bistatic angles and wind directions, is shown in Figure 12. The sea parameters were those from Section 2.

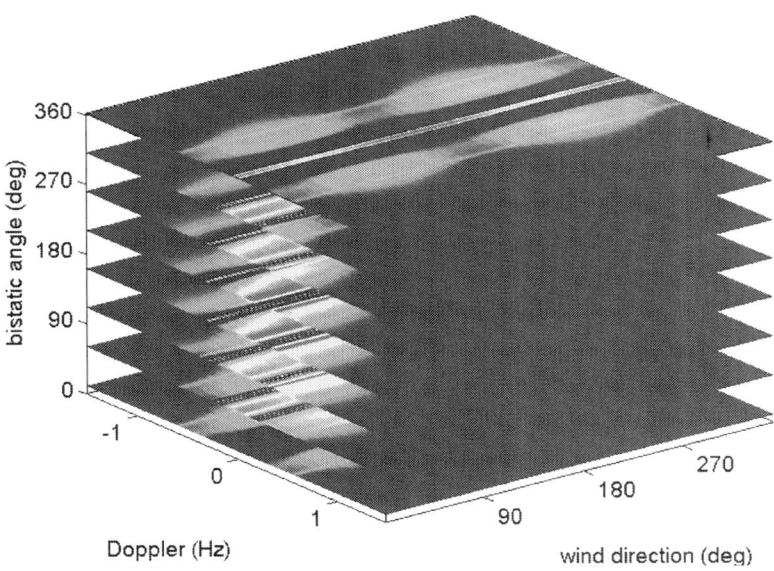

Figure 12: A small subset of the database of Doppler spectra evaluated for a particular wind speed and radar frequency, but with all combinations of wind direction and bistatic scattering angle.

For an operational deployment, one would compute figures of merit averaged over time of day and the seasons, for which we would need wind, wave and current climatologies. If appropriate, a weighting factor could be applied to effect diurnal or seasonal priorities.

For each of these figures of merit, the value lies in the interval $[0,1]$, increasing with the merit of the solution. Two simple options for the function to be minimised are $(1-OF)$ and OF^{-1}.

The figures of merit developed above apply to individual radars but the essence of the problem under consideration is optimisation of a network. The extension to the network case begins from the observation that, at any given moment, the target will be detected if at least one radar is able to achieve detection. For a set of radars operating in monostatic mode – what has been termed 'stereoscopic radar' (Anderson, 1990) – this can be encapsulated in the following expression:

$$OF_3 = \frac{1}{n}\sum_{k=1}^{n}\left[1 - \prod_{j=1}^{n}\left[1 - \max_{\omega \in Z} H\left(s_j(\omega;r) - c_j(\omega;r) - \varepsilon_j\right)\right]\right]$$

(8)

However, we need also to allow for bistatic detection, which has been shown (Anderson, 1990) to increase the probability of detection by circumventing the possibility of double blind speeds in stereoscopic configurations. This leads to

$$OF_4 = \frac{1}{n^2}\sum_{i=1}^{n}\sum_{j=1}^{n}\left[1 - \prod_{i=1}^{n}\prod_{j=1}^{n}\left[1 - \max_{\omega \in Z} H\left(s_{ij}(\omega;r) - c_{ij}(\omega;r) - \varepsilon_i\right)\right]\right]$$

(9)

While this formulation seems reasonable as far as detection is concerned, it does not take into account the advantage of detecting a target with two radars simultaneously, from different directions. Not only is the probability of detection increased but detection-to-track association is improved; this is an important consideration in the dense traffic environment of the South China Sea where ships are on average only ~ 10 km apart, not much more than the radar range resolution and less than the azimuthal resolution of the smaller radars. Accordingly, when two radars can view a region, we could take dual detectability into account via a performance enhancement factor which is a function of the angle subtended at the target by the two radars.

Current Mapping

In principle, surface current mapping is a relatively simple operation, relying as it does on two very strong peaks in the Doppler spectrum. A fairly rudimentary objective function is the predicted clutter-to-noise ratio, which can be defined for both monostatic and bistatic measurements. Given the length scales of fine structure in the current field in the South China Sea, as shown in Figure 6, some refinements are needed if the function is to serve its purpose effectively.

The most important is consideration of the phenomenon of geometric dilution of precision. The parameters which govern the GDOP for current measurement are (i) the bistatic angle 2β subtended by the two radar axes, and (ii) the crossing angle χ, that is the angle between the nominal current direction and the bisector of the two radar axes. These are indicated on Fig. 4. The theory of GDOP is widely reported (see for example Chapman et al, 1997, Emery et al, 2004) and will not be repeated here. If, instead of the total current vector, one is interested in the components along and perpendicular to a given direction, a slightly different function emerges; we are then interested in the component shown as $u\perp$, so the relevant crossing angle is φ and the error associated with GDOP must be computed using this angle.

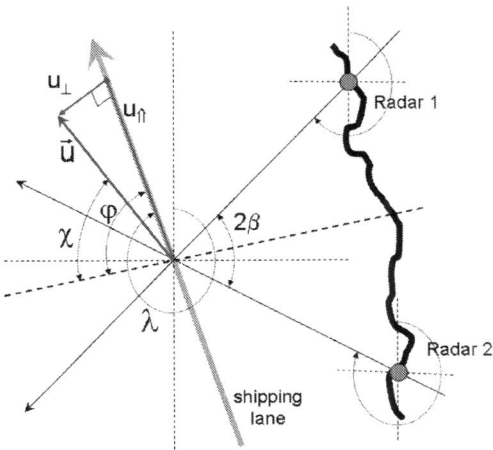

Figure 13: Geometry of bistatic illumination, with current vector and shipping lane.

Let the radial component of current velocity derived from a measurement by radar 1 be designated $u_1 + \varepsilon_1$, where ε_1 represents random measurement error, and similarly, that of radar 2 by $u_2 + \varepsilon_2$. The estimates of velocity parallel to and normal to the bisector axis are then given by

$$u_p = \frac{\left(u_1 + u_2 + \varepsilon_1 + \varepsilon_2\right)}{2\cos\beta}$$

(10)

$$u_n = \frac{\left(u_1 - u_2 + \varepsilon_1 - \varepsilon_2\right)}{2\sin\beta}$$

(11)

We transform this vector measurement into the coordinate system defined by the lane axis and its normal by means of a rotation:

$$\begin{bmatrix} u_\Uparrow \\ u_\perp \end{bmatrix} = \begin{bmatrix} \cos\varphi & \sin\varphi \\ -\sin\varphi & \cos\varphi \end{bmatrix} \begin{bmatrix} u_p \\ u_n \end{bmatrix}$$

(12)

Solving for $u_\perp$

$$u_\perp = \left(\frac{\cos\varphi}{2\sin\beta} - \frac{\sin\varphi}{2\cos\beta}\right)u_1 - \left(\frac{\cos\varphi}{2\sin\beta} + \frac{\sin\varphi}{2\cos\beta}\right)u_2$$
$$\pm \left[\left(\frac{-\sin\varphi}{2\cos\beta} + \frac{\cos\varphi}{2\sin\beta}\right)\varepsilon_1 - \left(\frac{\cos\varphi}{2\sin\beta} + \frac{\sin\varphi}{2\cos\beta}\right)\varepsilon_2\right]$$

(13)

Assuming ε1 and ε2 are independent and identically distributed, the rms error ε is found by squaring and averaging the error term,

$$\varepsilon = \frac{\left[\cos^2\left(\varphi + \beta\right) + \cos^2\left(\varphi - \beta\right)\right]^{1/2}}{\sin\beta}\varepsilon_1$$

(14)

The GDOP is defined as the ratio of the rms error ε to the error ε_1 associated with an individual radar. Numerical evaluation (Anderson, 2013) shows that the geometry has a major bearing on the accuracy of HFSWR current estimates, more than doubling the errors once the radars depart from orthogonal viewing geometry by more than 50°.

Visibility and Topographic Constraints

The figures of merit and associated objective functions developed in the preceding sections have made one assumption which demands explicit representation – the spatial integrations have made no allowance for blocking of the signal path from radar to patch of interest by an intervening land mass, either an island or part of the mainland. As it happens, HF surface waves can propagate across land, though with much greater attenuation than across sea, and there is an unusual effect (the Millington effect) through which a considerable fraction of signal strength is restored once the signal reaches the sea beyond the intervening land mass. Nevertheless, unless it cannot be avoided, it is better not to entertain the possibility of exploiting signals which have propagated across one or more islands. We can formalise this constraint on single site acceptability as follows.

Suppose there are K landmasses $\{D_j\}_{j=1,K}$ with coastlines $\{\partial D_j\}_{j=1,K}$ adjoining a sea or ocean of which a region W is to be monitored. From the k-th coastline, ∂D_k, construct ∂D_k^+ as follows:

$$\partial \tilde{D}_k^+ = \left\{ r \in \partial D_k \left| \begin{array}{l} \left\{ \alpha r + (1-\alpha) r' \right\} \cap D_m = \{0\} \\ \forall r' \in W, \\ \forall m = 1, K; \, \alpha \in [0,1] \end{array} \right. \right\}$$

(15)

Then $\partial D_K^+ \subset \partial$ Dk is the subset of the coastline of the k-th landmass which has an unobstructed view of the region W. If a radar is to be placed on Dk , then it must lie on ∂D_K^+.

In the present study, where each radar can hope to survey at most a part of the area of concern, it is more appropriate to assign the coverage arc at each candidate radar site and measure the effectiveness of that coverage according to the metrics defined earlier.

Accordingly we have chosen to define the coverage arcs by the requirement that they exclude any directions which meet regions W for which the site does not belong to

∂D_K^+.

For the present illustrative purposes we shall not impose other site-specific constraints such as conditions on local topography or coastline orientation and curvature, though these too could be added if desired.

RADAR NETWORKS FOR THE SOUTH CHINA SEA

The tools and procedures described in the preceding sections are of general applicability, but the success of the network optimisation relies on making best possible use of site-specific environmental information, not only oceanographic and meteorological but also the levels and patterns of HF noise. Often this information is unavailable or incomplete, but in most cases one can find climatological data which will serve adequately. We have used the South China Sea mainly as a context to illustrate the ways in which the geophysical information can be exploited, as well as the general issues that could drive network deployment. Needless to say, the South China Sea is of particular interest, so a number of network design experiments have been undertaken. They confirm the applicability of the genetic algorithm methodology to this context, reinforcing the conclusions of Anderson (2013) for the Strait of Malacca.

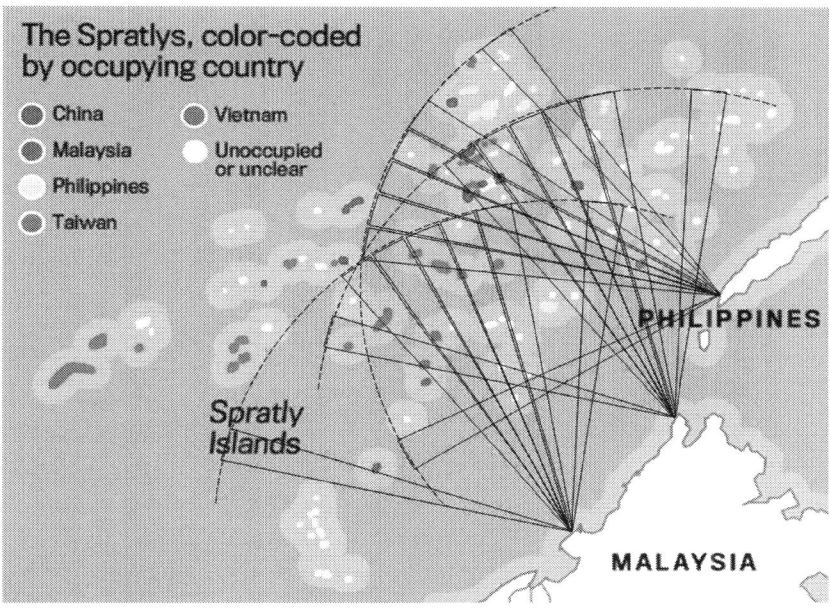

Figure 14: Part of an optimum solution, focussing on missions over the Spratly islands, showing the receive beam structure for a military-class phased array system.

An example of a candidate solution to a particular network optimisation problem is presented in Figure 14. It shows the coverage of the Spratly Islands afforded by a network of three military-class radars, addressing a combination of mission types. In this example, the optimality criterion of maximising performance for a fixed cost, discussed in Section 4.2, was used to constrain the solution space. A more comprehensive set of solutions, invoking a variety of optimality constraints, is in preparation (Anderson, 2014).

Note again that performance is a statistical quantity, with somewhat greater or perhaps far lesser coverage achieved on any given occasion.

CONCLUSIONS

The optimum deployment of a network of HFSWR systems is a highly complex task with many factors to be considered, especially when the radars are expected to perform multiple roles. Failure to treat the design problem with appropriate care could seriously degrade performance in one or more radar missions.

In this paper we have described a practical technique for HFSWR network design, based on a genetic algorithm adapted to multi-objective optimisation, and illustrated the method by placing it in the context of designing a multi-radar configuration system for remote sensing and surveillance of the South China Sea. The treatment pays particular attention to the construction of chromosomes and objective functions and extends previous measures of performance to allow for bistatic radar operations. In addition, we emphasise the importance of exploiting a priori knowledge about the regional geography, meteorology and oceanography in the design procedures.

A key advantage of the Pareto dominance formulation developed by Anderson (2013) and incorporated here is that it efficiently identifies those solutions which are superior to their fellows according to every criterion tested, and hence greatly reduces the range of design options which need to be considered. The designer is presented with a range of candidate optimal solutions which can be assessed according to additional considerations which may not be meaningfully quantifiable. When we consider domains such as the South China Sea, where complex geopolitical issues may arise, the virtues of this approach are self-evident.

REFERENCES

1.   Anderson, S.J. (1990). Stereoscopic and Bistatic Skywave Radars : Assessment of Capabilities and Limitations, Proceedings of Radarcon-90, Adelaide, Australia, 305-313.

2.   Anderson, S.J., Edwards, P.J., Marrone, P., and Abramovich, Y.I. (2003). Investigations with SECAR - a bistatic HF surface wave radar, Proceedings of IEEE International Conference on Radar, RADAR 2003, Adelaide.

3.   Anderson, S.J. (2013). Optimizing HF Radar Siting for Surveillance and Remote Sensing in the Strait of Malacca, IEEE Transactions on Geoscience and Remote Sensing, Vol.51, No.3, 1805-1816.

4.   Anderson, S.J., Darces, M., Helier, M., and Payet, N. (2013). Accelerated convergence of genetic algorithms for application to real-time inverse problems, Proceedings of the 4th Inverse Problems, Design and Optimization Syposium, IPDO-2013, Albi, France, 149-152.

5.   Anderson, S.J. (2014). HF radar network optimisation : Case studies for the South China Sea. In preparation.

6.   Barrick D E. (1998). EEZ Surveillance-The Compact HF Radar Alternative, EEZ Technology, Edition 3. London: ICG Publishing Ltd,. 125-129. See also the CODAR website http://www.codar. com/ and references therein for detailed information about the SeaSonde family of HFSWR systems.

7.   Chapman, R.D., Shay, L.K., Graber, H.C., Edson, J.B., Karachintsev, A., Trump, C.L., and Ross, D.B. (1997). On the accuracy of HF radar surface current measurements : Inter-comparisons with ship-base sensors, Journal of Geophysical Research, vol. 102, C8, 118737-18748.

8.   Chu, P.C., Qi, Y., Chen, Y., Shi, P., and Mao, Q. (2003). Validation of Wavewatch-III Using the TOPEX/POSEIDON Data, Proceedings of SPIE Conference on Remote Sensing of the Ocean and Sea Ice, Barcelona, Spain.

9.   Chu, P.C., Qi, Y., Chen, Y., and Mao, Q. (2004). South China Sea Wind-Wave Characteristics. Part I: Validation of Wavewatch-III Using the TOPEX/POSEIDON Data, Journal of Oceanic Technology, Vol.21, No.11, 1718-1733.

10.   Emery, B.M., Washburn, L., and Harlan, J.A. (2004). Evaluating radial current measurements from CODAR high-frequency radars with moored current meters, Journal of Atmospheric and Oceanic Technology, vol. 21, no. 8, pp. 1259-1271.

11.   Helzel, T., Kniephoff, M., and Pettersen, L. (2010). Oceanography radar system WERA : features, accuracy, reliability and limitations,

Turkish Journal of Electrical Engineering and Computer Science, vol. 18, no. 3, 389-397. See also the Helzel Messtechnik website http://www.helzel.com/ and references therein for detailed information about the WERA family of HFSWR systems.

12.   International Crisis Group. (2012a). Stirring up the South China Sea (I), International Crisis Group Asia Report No. 223.

13.   International Crisis Group. (2012b). Stirring up the South China Sea (II): Regional Responses, International Crisis Group Asia Report No. 229.

14.   Lipa, B., Isaacson, J., Nyden, B., and Barrick, D. (2012), Tsunami arrival detection with high frequency radar, Remote Sensing, Vol.4, No.11, 1448-1461.

15.   Marghany, M. (2009). Volterra - Lax-wendroff algorithm for modelling sea surface flow pattern fromjason-1 satellite altimeter data. Lecture Notes in Computer Science (including subseries Lecture Notes in Artificial Intelligence and Lecture Notes in Bioinformatics) Volume 5730 LNCS, 1-18 .

16.   Marghany, M. (2011). Developing robust model for retrieving sea surface current from RADARSAT-1 SAR satellite data, International Journal of the Physical Sciences. Vol. 6 (29), 6630-6637.

17.   Marghany, M. (2012). Three-Dimensional Coastal Front Visualization from RADARSAT-1 SAR Satellite Data. In Murgante B. et al. (eds.): Lecture Notes in Computer Science (ICCSA 2012), Part III, LNCS 7335, 447–456.

18.   Mirzaei, A., Tangang, F., Liew, J., Mustapha, M.A., Husain, M.L., and Akhir, F.A. (2013). Wave climate simulatrion for southern region of the South China Sea, Ocean Dynamics, Vol.63, 961-977.

19.   Ninh, P.V., Quynh, D.N., Lanh, V.V., and Lien, T.V. (2000). Geostrophic and drift current in the South China Sea, Area IV: Vietnamese waters, Proceedings of the SEADEC Seminar on Fishery Resources in the South China Sea, Area IV: Vietnamese Waters, 365-373.

20.   O'Rourke, R. (2013). Maritime Territorial and Exclusive Economic Zone (EEZ) Disputes Involving China: Issues for Congress, Congressional Research Service Report for Congress 7-5700 R42784.

21. Ponsford, A.M. (2012). Persistent surveillance of the 200 nautical mile Exclusive Economic Zone using Raytheon's land-based high frequency surface wave radar, Raytheon Technology Today, 2012, issue 2, 25-27.

22. Wang, J., Li, M., Liu, Y., Zhang, H., Zou, W., and Cheng, L. (2014). Safety assessment of shipping routes in the South China Sea based on the fuzzy analytic hierarchy process, Safety Science, Vol.62, 46-57.

Chapter 7

Studies on the Evaporation Regulation Mechanisms of Crude Oil and Petroleum Products

Merv F. Fingas

Spill Science, Edmonton, Canada

ABSTRACT

Various concepts for oil evaporation prediction are summarized. Models can be divided into those models that use the basis of air-boundary-regulation or those that do not. Experiments were conducted to determine if oil and petroleum evaporation is regulated by the saturation of the air boundary layer. Experiments included the examination of the evaporation rate with and without wind, in which case it was found that evaporation rates were similar for all wind conditions and no-wind conditions. Experiments where the area and mass varied showed that boundary-layer regulation was not governing for petroleum products. Under all experimental and environmental conditions, oils or petroleum products were not found to be boundary-layer regulated. Experiments on the rate of evaporation of pure compounds showed that compounds larger than Decane were not boundary-layer regulated. Many oils and petroleum products contain few compounds smaller than decane, and this explains why their evaporation is not air boundary-layer limited. Comparison of the air saturation levels of various oils and petroleum products shows that the saturation concentration of water, which is strongly air boundary-regulated, is significantly less than that of several petroleum hydrocarbons. Lack of air boundary-layer regulation for oils is shown to be a result of both this higher saturation concentration as well as a low (below boundary-layer value) evaporation rate.

INTRODUCTION

Evaporation is an important process for most oil spills. In a few days, typical crude oils can lose up to 45% of their volume. The Macondo oil lost up to 60% in a short time when released under water at high pressure [1]. Almost all oil spill models include evaporation as a process and output of the model. Evaporation plays a prime role in the fate of most oils. Many crude oils must undergo evaporation before they will form water-in-oil emulsions [1]. Light oils will change very dramatically from fluid to viscous. Heavy oils will become solid-like. Many oils after long evaporative exposure form tar balls or heavy tar mats. Despite the importance of the process, little work has been conducted on the basic physics and chemistry of oil spill evaporation [2]. The difficulty with studying oil evaporation is that oil is a mixture of hundreds of compounds and oil composition varies from source to source and even over time. Much of the work described in the older literature focuses on calibrating equations developed for water evaporation [2].

The mechanisms that regulate evaporation are important [3, 4]. Evaporation of a liquid can be considered as the movement of molecules from the surface into the vapour phase above it. The immediate layer of air above the evaporation surface is known as the air boundary layer [5]. This boundary layer is the intermediate interface between the air and the liquid and might be viewed as very thin such as less than one mm. The characteristics of this air boundary layer can influence evaporation.

In the case of water, the boundary layer regulates the evaporation rate. Air can hold a variable amount of water, depending on temperature, as expressed by the relative humidity. Under conditions where the air boundary layer is not moving (no wind) or has low turbulence, the air immediately above the water quickly becomes saturated and evaporation slows. The actual evaporation of water proceeds at a small fraction of the possible evaporation rate because of the saturation of the boundary layer. The air-boundary-layer physics is then said to regulate the evaporation of water. This regulation manifests as the increase of evaporation with wind or turbulence. When turbulence is weak, evaporation can slow down by orders-of-magnitude. The molecular diffusion of water molecules through air is at least 10^3 times slower than turbulent diffusion [5]. If the evaporation of oil was like that of water and was air boundary-layer regulated, one could write the mass transfer rate in semi-empirical form (also in generic and unitless form) as:

$$E = KCT_u S$$

(1)

where E is the evaporation rate in mass per unit area, K is the mass transfer rate of the evaporating liquid, presumed constant for a given set of physical conditions, sometimes denoted as k_g (gas phase mass transfer coefficient, which may incorporate some of the other parameters noted here), C is the concentration (mass) of the evaporating fluid as a mass per volume, T_u is a factor characterizing the relative intensity of turbulence, S is a factor that relates to the saturation of the boundary layer above the evaporating liquid. The saturation parameter, S, represents the effects of local advection on saturation dynamics. If the air is already saturated with the compound in question, the evaporation rate approaches zero. This also relates to the scale length of an evaporating pool. If one views a large pool over which a wind is blowing, there is a high probability that the air is saturated downwind and the evaporation rate per unit area is lower than for a smaller pool. It should be noted that there are many equivalent ways of expressing this fundamental evaporation equation.

Much of the pioneering work for water evaporation work was performed by Sutton [6]. Sutton proposed the following equation based largely on empirical work:

$$E = KC_S U^{7/9} d^{-1/9} Sc^{-r}$$

(2)

where C_s is the concentration of the evaporating fluid (mass/volume), U is the wind speed, d is the area of the pool, Sc is the Schmidt number and r is the empirical exponent assigned values from 0 to 2/3. Other parameters are defined as above. The terms in this equation are analogous to the very generic equation (1), proposed above. The turbulence is expressed by a combination of the wind speed, U, and the Schmidt number, Sc. The Schmidt number is the ratio of kinematic viscosity of air ( ) to the molecular diffusivity (D) of the diffusing gas in air, i.e. a dimensionless expression of the molecular diffusivity of the evaporating substance in air. The coefficient of the wind power typifies the turbulence level. The value of 0.78 (7/9) as chosen by Sutton, represents a turbulent wind whereas a coefficient of 0.5 would represent a wind flow that was more laminar. The scale length is represented by d and has been given an empirical exponent of −1/9. This represents, for water, a weak dependence on size. The exponent of the Schmidt number, r, represents the effect of the diffusivity of the particular chemical, and historically was assigned values between 0 and 2/3 [5].

This expression for water evaporation was subsequently used by those working on oil spills to predict and describe oil and petroleum evaporation. Much of the literature follows the work of Mackay [7, 8]. Mackay and Matsugu [7] corrected the equations for hydrocarbons using the evaporation rate of cumene. Data on the evaporation of water and cumene have been used to correlate the gas phase mass transfer coefficient as a function of wind-speed and pool size by the equation:

$$K_m = 0.0292 U^{0.78} X^{-0.11} Sc^{-0.67}$$

(3)

Where K_m is the mass transfer coefficient in units of mass per unit time and X is the pool diameter or the scale size of evaporating area. Stiver and Mackay [8] subsequently developed this further by adding a second equation:

$$N = k_m A P / (RT)$$

(4)

Where N is the evaporative molar flux (mol/s), k_m is the mass transfer coefficient at the prevailing wind (m/s), A is the area (m^2), P is the vapour pressure of the bulk liquid (Pascals), R is the gas constant [8.314 Joules/ (mol-K)], and T is the temperature (K).

Thus, air boundary layer regulation was assumed to be the primary regulation mechanism for oil and petroleum evaporation. This assumption was never tested by experimentation, as revealed by a literature search [2]. The implications of these assumptions are that evaporation rate for a given oil is increased by:

*   increasing turbulence
*   increasing wind speed, and
*   increasing the surface area of a given mass of oil.

These factors can then be verified experimentally to test if oil is boundary-layer regulated or not. These factors formed the basis of experimentation for this paper.

EXPERIMENTAL

Evaporation rate was measured by weight loss using an electronic balance. The balance was a Mettler PM4000. The weight was recorded using a laptop computer, a serial cable to the balance and the software program, "Collect" (Labtronics, Richmond, Ontario).

Measurements were conducted in the following fashion. A tared petri dish of defined size was loaded with a measured amount of oil. At the end of the experiment vessels were cleaned and rinsed with dichloromethane and a new experiment started. The weight loss dishes were standard glass petri dishes from Corning. A standard 139 mm diameter (ID) dish was used for most experiments. For the experiments in which area was a variable, dishes of other diameters were employed. Diameters and other dimensions were measured using a Mitutoyo digital vernier caliper. The lip, height of the dish above the oil, with the 139 mm dish varied from 2 to 10 mm depending on depth of fill. For the other dishes the lip varied from 2 to 20 mm.

Measurements were done in one of three locations; inside a fume hood, inside a controlled temperature room, or on a counter top. Some experiments were conducted in the fume hood, where there was no temperature regulation. Temperatures were measured using a Keithley 871 digital thermometer with a thermocouple supplied by the same firm. Temperatures were taken at the beginning and the end of a given experimental run.

The constant temperature chamber (room) employed was a Constant Temperature model. It could maintain temperatures from −40°C to +60°C and regulate the chosen temperature within ±1°C.

In experiments involving wind, air velocities were measured using a Taylor vane anemometer and a Tadi, "Digital Pocket Anemometer". Measurements were taken at the closest position above the glass vessel floor and at the lip level. These velocities were later confirmed using a hot wire anemometer and appropriate data manipulations of the outputs. The anemometer was a TSI— Thermo Systems model 1053b, with power supply (TSI model 1051-1), averaging circuit (TSI model 1047) and signal linearlizing circuit (TSI model 1052). The voltage from the averaging circuit was read with a Fluke 1053 voltmeter. The hot wire sensor (TSI model 1213-60) was angled at 45°. The sensor probe resistance at 0°C was 7.21 ohms and the sensor was operated at 12 ohms for a recommended operating temperature of 250°C. Data from the hot wire anemometer was collected on a Campbell Scientific CR-10 data logger at a rate of 64 Hz.

Evaporation data were collected on a laptop computer and subsequently transferred to other computers for analysis. The "Collect" program records time and the weight directly. Data were recorded in ASCII format and converted to Excel format. Curve fitting was performed using the software program "TableCurve", Jandel Scientific Corporation, San Raphael, California.

Oils were taken from supplies of Environment Canada and were supplied by various oil companies for environmental testing. Table 1 lists the properties and descriptions of the test liquids [9].

RESULTS AND DISCUSSION

Table 2 lists the experiments performed and the results in terms of the best fit equations. These were done by curve fitting using the program

Table Curve, as noted above. The best fit was done on the basis of the simplest equation fitting with the highest regression coefficient (R^2). The results are presented in the order of the experimental series:

Wind Experiments

Experiments on the evaporation of oil with and without wind, were conducted with three oils, ASMB (Alberta Sweet Mixed Blend crude oil), Gasoline, FCC Heavy Cycle (a processed oil), and with water. Water formed a baseline data set since much is known about its evaporation behaviour [3, 4]. Regressions on the data were performed and the equation parameters calculated, are shown in Table 3. Curve coefficients are the constants from the best fit equation (Evap = a ln (t),

t = time in minutes, for logarithmic equations or Evap = a$\sqrt{t}$, for the square root equations). Data were calculated separately for percentage of weight lost and absolute weight. Both values show the small relative upward tendency with respect to wind effects. The plots of wind speed versus the evaporation rate (as a percentage of weight lost) for each oil type are shown in Figures 1 to 4. These figures show that the evaporation rates for oils and even the light products, gasoline and FCC Heavy Cycle, are not increased by a significant amount with increasing wind speed. In some cases, there is a small rise from the 0-wind level to the 1-m/s level, but after that, the rate remains relatively constant. The evaporation rate after the 0-wind value is nearly identical for all oils. The oil evaporation data can be compared to the evaporation of water, as illustrated in Figure 4.

Table 1: Properties of the test liquids

Test Liquid	Description	Density g/mL	Boiling Point °C
ASMB	Alberta Sweet Mixed Blend—A common crude oil in Canada	0.839	initial-37
Water		1	100
FCC-heavy	A highly-cycled refinery intermediate containing few components	0.908	

Gasoline	Standard automotive gasoline	0.709	initial-5
Benzene	Pure Hydrocarbon C6	0.879	80.1
Dodecane	Pure Hydrocarbon C10	0.749	213
Undecane	Pure Hydrocarbon Cl I	0.742	196
p-Xylene	Pure Hydrocarbon C8	0.861	139
Nonane	Pure Hydrocarbon C9	0.722	151
Decane	Pure Hydrocarbon C10	0.73	174
Heptane	Pure Hydrocarbon C7	0.684	98
Octane	Pure Hydrocarbon C8	0.703	126
Decahydron	Decahydronaphthalene pure hydrocarbon CIO	0.893	195
Tridecane	Pure Hydrocarbon C13	0.755	226
Hexadecane	Pure Hydrocarbon C16	0.773	287

Table 2: Experimental summary

Number	Experimental Purpose	Oil Type	Total Time (hr)	Pan (cm²) Area	Initial (mm) Thickness	Temp °C	Wind m/s	Variable	Variable Value	R^2 Best Equation	Best Equation
1	Thickness	ASMB	15	151	0.65	21.2	0	thick	0.65	0.991	ln
2	Thickness	ASMB	22	268	0.72	21	0	thick	0.72	0.978	ln
3	Thickness	ASMB	23	270	1.3	21.8	0	thick	1.3	0.97	ln
4	Thickness	ASMB	182	151	0.63	22.6	0	thick	0.63	0.99	ln
5	Thickness	ASMB	15	151	1.59	22.4	0	thick	1.59	0.937	ln
6	Thickness	ASMB	51	151	1.78	21.9	0	thick	1.78	0.975	ln
7	Thickness	ASMB	65	151	2.14	24.4	0	thick	2.14	0.954	ln
8	Thickness	ASMB	25	151	2.69	23.8	0	thick	2.69	0.952	ln
9	Thickness	ASMB	73	151	2.84	21.7	0	thick	2.84	0.96	ln
10	Thickness	ASMB	36	151	4.55	22.8	0	thick	4.55	0.963	ln
11	Thickness	ASMB	18	151	9.08	20.1	0	thick	9.08	0.879	ln
12	Thickness	ASMB	73	151	7.61	20.3	0	thick	7.61	0.886	ln
13	Thickness	ASMB	217	151	5.21	20	0	thick	5.21	0.937	ln
14	Thickness	ASMB	64	151	1.53	22.1	0	thick	1.53	0.981	ln
15	Thickness	ASMB	56	151	3.21	17.8	0	thick	3.21	0.952	ln
16	Thickness	ASMB	47	151	1.33	19.2	0	thick	1.33	0.987	ln
17	Thickness	ASMB	23	151	0.59	18.8	0	thick	0.59	0.988	ln
18	Thickness	ASMB	25	151	0.63	20.1	0	thick	0.63	0.985	ln
19	Thickness	ASMB	71	151	1.96	23.1	0	thick	1.96	0.976	ln
20	Thickness	ASMB	32	151	2.54	18.6	0	thick	2.54	0.977	ln

21	Thickness	ASMB	89	151	5.27	22.9	0	thick	5.27	0.98	ln
22	Thickness	ASMB	76	151	1.43	20.4	0	thick	1.43	0.993	ln
23	Thickness	ASMB	66	151	1.39	20.3	0	thick	1.39	0.986	ln
24	Thickness	ASMB	88	151	2.8	19.1	0	thick	2.8	0.962	ln
25	Area	ASMB	50	16	7.45	24.2	0	area	16 cm^2	0.969	ln
26	Area	ASMB	25	16	3.72	23.9	0	area	16 cm^2	0.96	ln
27	Area	ASMB	21	16	1.58	8	0	area	16 cm^2	0.72	ln
28	Area	ASMB	25	16	0.79	24.6	0	area	16 cm^2	0.791	ln
29	Area	ASMB	50	62	3.84	22.5	0	area	62 cm^2	0.992	ln
30	Area	ASMB	22	62	1.92	15.6	0	area	62 cm^2	0.996	ln
31	Area	ASMB	26	62	1.58	25.3	0	area	62 cm^2	0.982	ln
32	Area	ASMB	23	62	0.79	23.8	0	area	62 cm^2	0.994	ln
33	Area	ASMB	24	161	1.48	21	0	area	161 cm^2	0.987	ln
34	Area	ASMB	23	161	0.79	25.2	0	area	161 cm^2	0.973	ln
35	Area	ASMB	50	161	1.58	23.9	0	area	161 cm^2	0.941	ln
36	Area	ASMB	83	161	3.7	19.1	0	area	161 cm^2	0.933	ln
37	Area	ASMB	50	161	2.22	21	0	area	161 cm^2	0.99	ln
38	Area	ASMB	25	161	0.74	20	0	area	161 cm^2	0.953	ln
39	Area	ASMB	74	206	1.58	18	0	area	206 cm^2	0.984	ln
40	Area	ASMB	20	206	0.79	21	0	area	206 cm^2	0.974	ln
41	Area	ASMB	51	206	1.16	19.5	0	area	206 cm^2	0.963	ln
42	Area	ASMB	44	151	1.58	20.5	0	area	151 cm^2	0.993	ln
43	Area	ASMB	26	151	0.79	19	0	area	151 cm^2	0.994	ln
44	Wind	ASMB	23	151	1.58	22.9	1.45	wind	1.0 m/s	0.98	ln
45	Wind	ASMB	24	151	1.58	22	1.45	wind	1.0 m/s	0.972	ln
46	Wind	ASMB	42	151	3.16	21.1	1.45	wind	1.0 m/s	0.99	ln
47	Wind	ASMB	46	151	3.16	21.2	1.45	wind	1.0 m/s	0.993	ln
48	Wind	Water	3	151	1.32	21.8	1.45	wind	1.0 m/s	0.997	lin
49	Wind	Water	3	151	1.32	21.8	1.45	wind	1.0 m/s	0.997	lin
50	Wind	Water	3	151	2.65	21.8	1.45	wind	1.0 m/s	0.999	lin
51	Wind	ASMB	21	151	1.58	22.1	1.65	wind	1.6 m/s	0.981	ln
52	Wind	ASMB	22	151	1.58	21.4	1.65	wind	1.6 m/s	0.949	ln
53	Wind	ASMB	23	151	1.58	21.4	1.65	wind	1.6 m/s	0.996	ln
54	Wind	ASMB	46	151	3.16	22.7	1.65	wind	1.6 m/s	0.986	ln
55	Wind	ASMB	20	151	1.58	22.8	1.65	wind	1.6 m/s	0.977	ln

56	Wind	Water	1	151	1.32	21.7	1.65	wind	1.6 m/s	0.998	lin
57	Wind	ASMB	17	151	1.58	23.9	1.65	wind	1.6 m/s	0.978	ln
58	Wind	Water	3	151	1.32	22.2	1.65	wind	1.6 m/s	0.999	lin
59	Wind	Water	5	151	2.65	23.6	1.65	wind	1.6 m/s	0.989	lin
60	Wind	ASMB	22	151	1.58	24.3	1.65	wind	1.6 m/s	0.981	ln
61	Wind	Water	1	151	1.32	23.4	1.85	wind	2.1 m/s	0.998	lin
62	Wind	ASMB	44	151	3.16	23	1.85	wind	2.1 m/s	0.991	ln
63	Wind	ASMB	6	151	1.58	21.7	1.85	wind	2.1 m/s	0.993	ln
64	Wind	ASMB	39	151	3.16	20.4	1.85	wind	2.1 m/s	0.993	ln
65	Wind	Water	2	151	1.32	21.8	1.85	wind	2.1 m/s	0.994	lin
66	Wind	Water	5	151	2.65	22.6	1.85	wind	2.1 m/s	0.998	lin
67	Wind	ASMB	12	151	1.58	22.4	1.85	wind	2.1 m/s	0.993	ln
68	Wind	FCC-heavy	32	151	2.92	21.7	1.85	wind	2.1 m/s	0.987	sq. rt.
69	Wind	Gasoline	1	151	1.87	22.6	1.85	wind	2.1 m/s	0.983	ln
70	Wind	Gasoline	2	151	3.74	22.4	1.85	wind	2.1 m/s	0.975	ln
71	Wind	FCC-heavy	22	151	1.46	22.3	1.85	wind	2.1 m/s	0.996	sq. rt.
72	Wind	ASMB	21	151	1.58	23.4	3.8	wind	2.5 m/s	0.981	ln
73	Wind	Water	1	151	1.32	22.4	3.8	wind	2.5 m/s	0.997	lin
74	Wind	Water	2	151	2.65	22.2	3.8	wind	2.5 m/s	0.999	lin
75	Wind	Gasoline	0	151	1.87	22.2	3.8	wind	2.5 m/s	0.984	ln
76	Wind	Gasoline	1	151	3.74	21.9	3.8	wind	2.5 m/s	0.994	ln
77	Wind	Water	3	151	1.32	21.7	0	wind	0	0.999	lin
78	Wind	FCC-heavy	47	151	2.92	21.4	3.8	wind	2.5 m/s	0.994	sq. rt.
79	Wind	FCC-heavy	39	151	1.46	22	3.8	wind	2.5 m/s	0.997	sq. rt.
80	Wind	ASMB	34	151	1.58	22.5	3.8	wind	2.5 m/s	0.993	ln
81	Wind	ASMB	18	151	3.16	21	3.8	wind	2.5 m/s	0.997	ln
82	Wind	Water	1	151	1.32	22	3.8	wind	2.5 m/s	0.986	lin
83	Wind	Water	2	151	2.65	22.9	3.8	wind	2.5 m/s	0.994	lin
84	Wind	FCC-heavy	19	151	1.46	23	3.8	wind	2.5 m/s	0.992	sq. rt.
85	Wind	Gasoline	1	151	1.87	22.1	1.65	wind	1.6 m/s	0.996	ln
86	Wind	Gasoline	3	151	3.74	22.4	1.65	wind	1.6 m/s	0.983	ln
87	Wind	FCC-heavy	40	151	2.92	22.3	1.65	wind	1.6 m/s	0.997	sq. rt.
88	Wind	Gasoline	1	151	1.87	21.8	1.45	wind	1.0 m/s	0.992	ln
89	Wind	Gasoline	2	151	3.74	22.1	1.45	wind	1.0 m/s	0.973	ln
90	Wind	FCC heavy	21	151	1.46	23.1	1.45	wind	1.0 m/s	0.99	sq. rt.

91	Wind	FCC heavy	51	151	2.92	24.2	1.45	wind	1.0 m/s	0.996	sq. rt.
92	Wind	FCC heavy	46	151	1.46	24	0	wind	0	0.986	sq. rt.
93	Wind	Water	3	151	1.32	23.9	0	wind	0	0.999	lin
94	Wind	FCC heavy	87	151	2.92	23.9	0	wind	0	0.996	ln
95	Wind	Water	8	151	2.65	25	0	wind	0	0.999	lin
96	Wind	Water	16	151	2.65	25.1	0	wind	0	0.998	lin
97	Wind	Gasoline	7	151	1.87	22.5	0	wind	0	0.92	ln
98	Wind	Gasoline	17	151	3.74	22.5	0	wind	0	0.944	ln
99	Wind	Water	6	151	1.32	23	0	wind	0	0.99	lin
100	Pure cmpd.	Benzene	2	151	1.51	23.9	0	rate		0.999	lin
101	Pure cmpd.	Dodecane	45	151	1.77	23.3	0	rate		0.999	lin
102	Pure cmpd.	Undecane	46	151	1.79	24.3	0	rate		0.999	lin
103	Pure cmpd.	p-Xylene	7	151	1.54	24	0	rate		0.989	lin
104	Pure cmpd.	Nonane	11	151	1.83	24	0	rate		0.999	lin
105	Pure cmpd.	Decane	19	151	1.81	22.3	0	rate		0.998	lin
106	Pure cmpd.	Heptane	3	151	1.94	18.5	0	rate		0.999	lin
107	Pure cmpd.	Octane	3	151	1.88	20.4	0	rate		0.997	lin
108	Pure cmpd.	Decahydronapthalene	18	151	1.48	21	0	rate		0.996	lin
109	Pure cmpd.	Tridecane	23	151	1.79	21.1	0	rate		0.986	lin
110	Pure cmpd.	Hexadecane	167	151	1.71	15	0	rate		0.847	lin

Table 3: Data from the wind tests

Type	Loading grams	Curve Coefficients*		Wind m/s	Type	Loading grams	Curve Coefficients		Wind m/s
		% evap	Abs. Wt.				% evap	Abs. Wt.	
ASMB	20	4.22	0.844	0	FCC heavy**	20	0.414	0.117	0
ASMB	20	5.28	1.06	1	FCC heavy	20	0.887	0.178	1
ASMB	20	5.3	1.06	1	FCC-heavy	20	0.8	0.161	2.1
ASMB	20	5.19	1.04	1.6	FCC-heavy	20	1.13	0.225	2.5
ASMB	20	5.27	1.05	1.6	FCC-heavy	20	0.905	0.181	2.5
ASMB	20	5.15	1.03	1.6					
ASMB	20	5.63	1.13	1.6	FCC heavy	20	0.414	0.2	0
ASMB	20	5.47	1.09	1.6	FCC heavy	40	0.66	0.264	1
ASMB	20	5.54	1.11	1.6	FCC-heavy	40	0.669	0.268	1.6
ASMB	20	5.78	1.16	2.1	FCC-heavy	40	0.557	0.223	2.1
ASMB	20	5.52	1.11	2.1	FCC-heavy	40	0.785	0.314	2.5
ASMB	20	5.82	1.16	2.5					
ASMB	20	5.52	1.1	2.5	Gasoline	20	12.2	3.36	0
					Gasoline	20	19.5	3.9	1
ASMB	40	4.09	2	0	Gasoline	20	19.7	3.93	1.6
ASMB	40	4.77	1.91	1	Gasoline	20	18.2	3.64	2.1
ASMB	40	4.77	1.91	1	Gasoline	20	21.6	4.32	2.5
ASMB	40	4.9	1.96	1.6					
ASMB	40	4.85	1.94	2.1	Gasoline	40	12.2	6	0

ASMB	40	4.99	2	2.1	Gasoline	40	16	6.4	1
ASMB	40	5.21	2.08	2.5	Gasoline	40	16.6	6.65	1.6
					Gasoline	40	15.4	6.15	2.1
Water	20	0.186	0.0372	0	Gasoline	40	16.6	6.64	2.5
Water	20	0.179	0.0357	0					
Water	20	0.178	0.0356	0	Water	40	0.088	0.0354	0
Water	20	0.592	0.118	1	Water	40	0.0778	0.0311	0
Water	20	0.612	0.112	1	Water	40	0.34	0.136	1
Water	20	0.512	0.102	1.6	Water	40	0.312	0.137	1.6
Water	20	0.515	0.103	1.6	Water	40	0.316	0.127	2.1
Water	20	0.7	0.14	2.1	Water	40	0.56	0.224	2.5
Water	20	0.603	0.12	2.1	Water	40	0.602	0.241	2.5
Water	20	1.02	0.206	2.5					
Water	20	1.04	0.209	2.5					

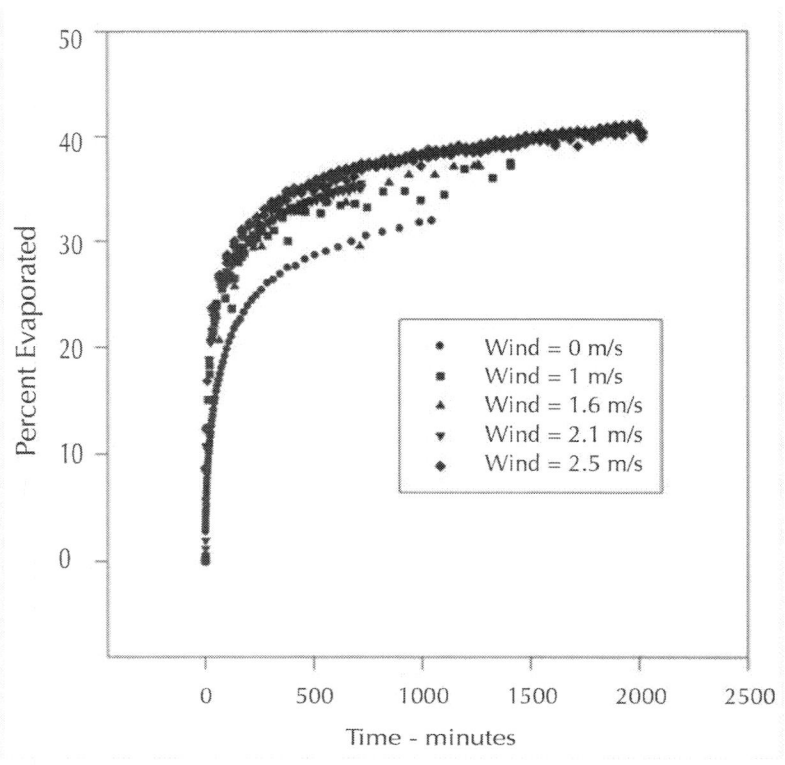

Figure 1: Evaporation of ASMB with varying wind velocities.

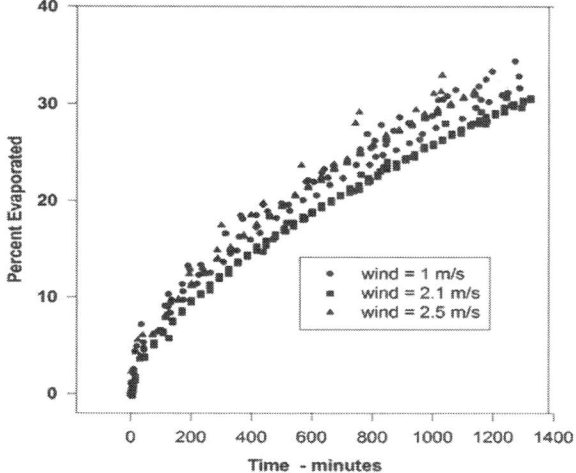

Figure 2: Evaporation of FCC-Heavy with varying wind velocities.

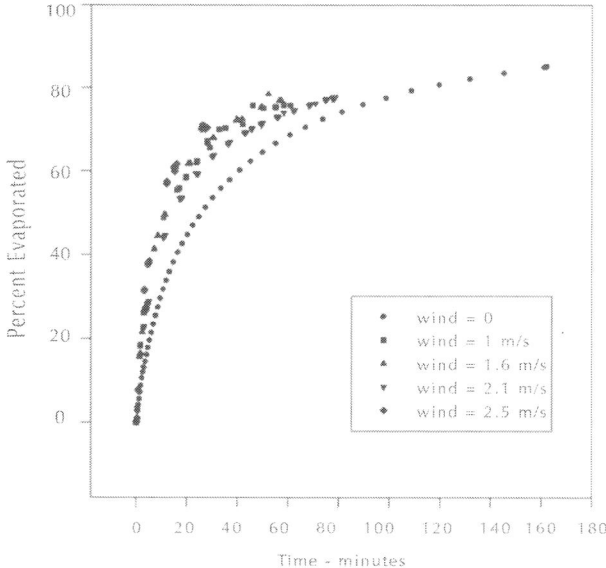

Figure 3: Evaporation of gasoline with varying wind velocities.

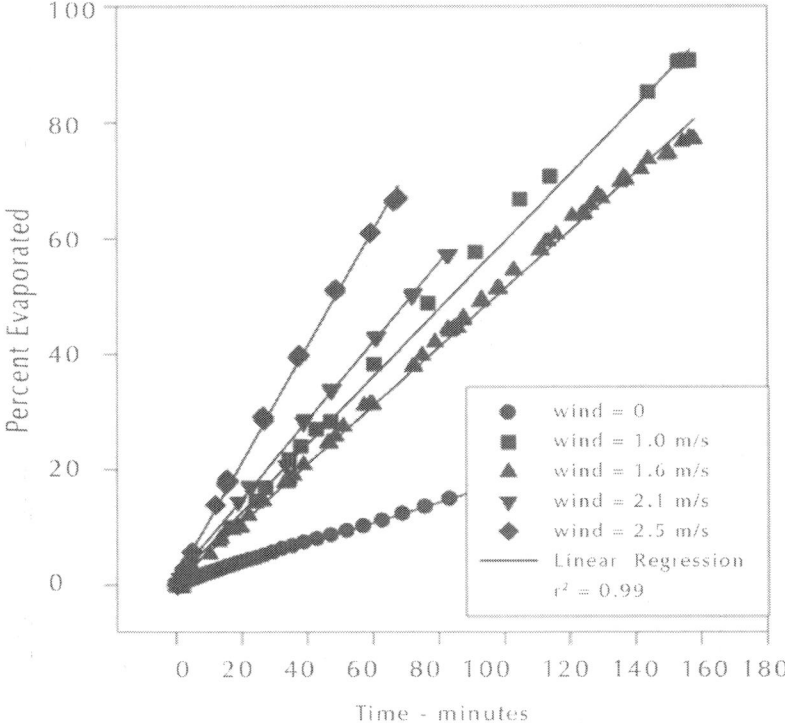

Figure 4: Evaporation of water (20 g) with varying wind velocities.

These data show the classical relationship of the water evaporation rate correlated with the wind speed (evaporation varies as $U^{0.78}$, where U is wind speed). This indicates that the oils used here are not boundary-layer regulated. Figure 5 shows the rates of evaporation compared to the wind speed for all the liquids used in this study. This figure shows the evaporation rates of all test liquids versus wind speed. The lines shown are those calculated by linear regression using the graphics software, SigmaPlot (Washington, DC). This clearly shows that water evaporation rate increased, as expected, with increasing wind velocity. The oils. ASMB, FCC heavy cycle and gasoline, do not show a measurable increase with increasing wind speed. In any case, the oils do not show the $U^{0.78}$ relationship that water shows.

All the above data show that oil is not boundary-layer regulated. Water shows the classic boundary-layer regulation.

Study of Mass and Evaporation Rate

ASMB oil was again used to conduct a series of experiments with volume as the major variant. Alternatively thickness and area were held constant to ensure that the strict relationship between these two variables did not affect the final regression results. Figure 6 illustrates the relationship between evaporation rate and volume of evaporation material (also equivalent to mass of evaporating material). This figure illustrates a strong correlation between oil mass (or volume) and evaporation rate. This suggests no air boundary-layer regulation is at work, since for an air boundary-layer regulated material evaporation is not affected by mass in the same area.

Study of the Evaporation of Pure Hydrocarbons—with and without Wind

A study of the evaporation rate of pure hydrocarbons was conducted to test the classic boundary-layer evaporation theory as applied to the hydrocarbon constituents of oils. The evaporation rate data are illustrated in Figure 7. This figure shows that the evaporation rates of the pure hydrocarbons have a variable response to wind. Heptane (hydrocarbon number 7) shows a large difference between evaporation rate in wind and no wind conditions, indicating boundary-layer regulation. Decane (carbon number 10) shows a lesser effect and Hexadecane (carbon number 16) shows a negligible difference between the two experimental conditions. This experiment shows the extent of boundary-regulation and the reason for the small or negligible degree of boundary-regulation shown by crude oils and petroleum products. Crude oil contains very little material with carbon numbers less than decane, often less than 3% of its composition [9]. Even the more volatile petroleum products, gasoline and diesel fuel only have limited amounts of compounds more volatile than decane, and thus are also not strongly boundary-layer regulated.

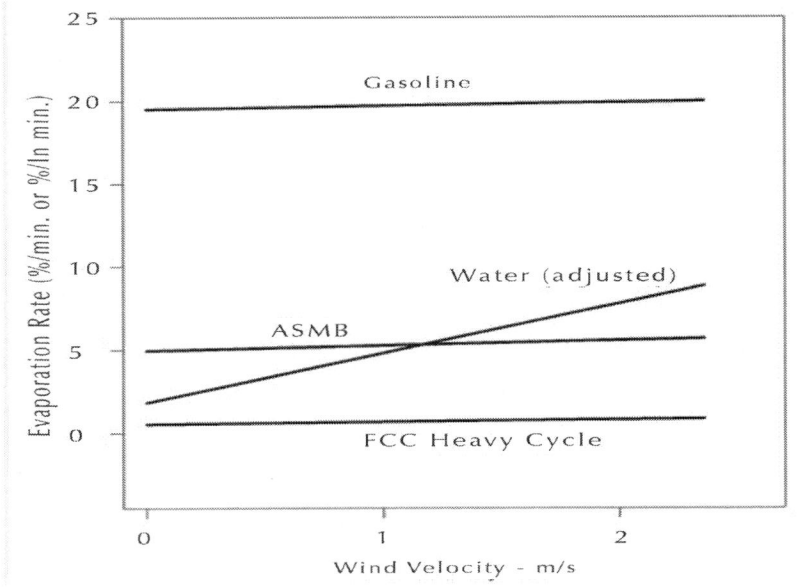

Figure 5: Correlation evaporation rates and wind velocities.

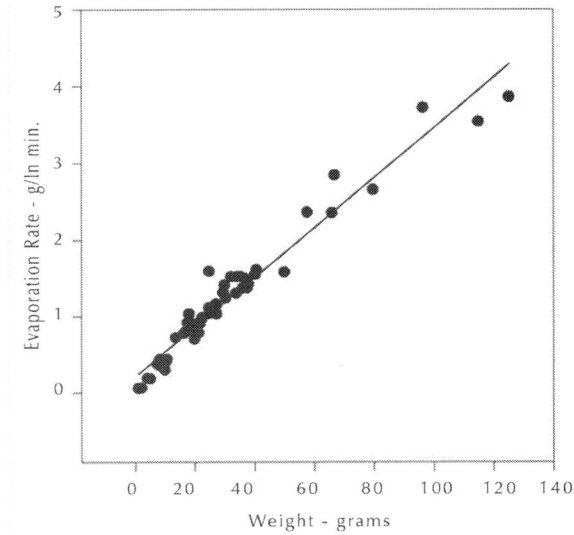

Figure 6: Correlation of mass with evaporation rate.

Saturation Concentration

Another evaluation of evaporation regulation is that of saturation concentration, the maximum concentration soluble in air. Table 4 lists the saturation concentrations of water and several oil components [10]. This table shows that saturation concentration of water is less than that of many common oil components. The saturation concentration of water is in fact, about two orders less in magnitude than the saturation concentration of volatile oil components such as pentane. This further explains why oil has a air boundary-layer limitation much higher than that of water and thus is not air boundary-layer regulated.

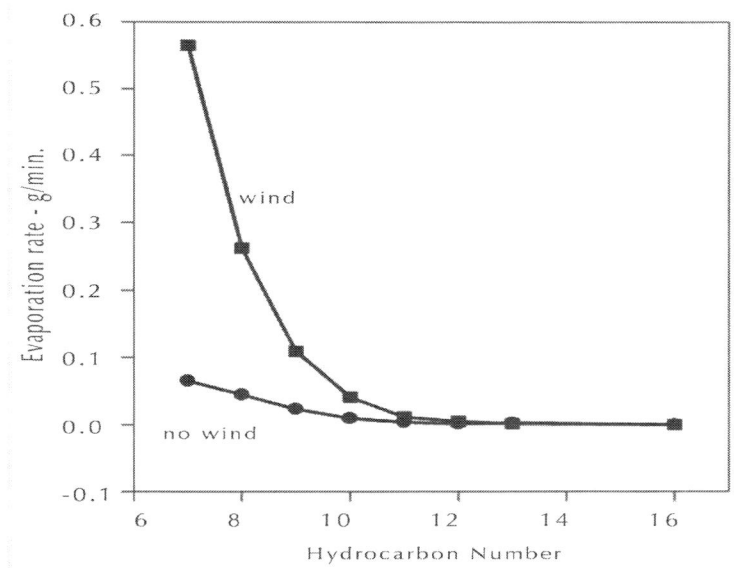

Figure 7: Evaporation rates of pure compounds.

Table 4: Saturation concentration of water and hydrocarbons

Substance	Saturation Concentration* in g/m3 at 25°C
water	20
n-pentane	1689

hexane	564
cyclohexane	357
benzene	319
n-heptane	196
methylcyclohexane	192
toluene	110
ethybenzene	40
p-xylene	38
m-xylene	35
o-xylene	29

*Values taken from Ullman's Encyclopedia [10].

CONCLUSIONS

Oil evaporation is not air boundary-layer regulated. The results of the following experimental series have shown the lack of boundary-layer regulation: 1) a study of the evaporation rate of several oils with increasing wind speed shows that the evaporation rate does not change measurably with wind level. Water, known to be boundary-layer regulated, does show a significant increase with wind speed, U (U^x, where x varies from 0.5 to 0.78, depending on the turbulence level); 2) the volume or mass of oil evaporating correlates with the evaporation rate. This is a strong indicator of the lack of boundary-layer regulation because with water, volume (rather than area) and rate do not correlate; 3) evaporation of pure hydrocarbons with and without wind (turbulence) shows that compounds larger than nonane and decane are not boundary-layer regulated. Most oil and hydrocarbon products consist of compounds larger than these two and thus would not be expected to be boundary-layer regulated.

Having concluded that boundary-layer regulation is not specifically applicable to oil evaporation, it remains to explain why this is so. The reason is twofold: oil evaporation is relatively slow compared to the threshold where it would be air boundary-layer regulated; and the threshold to boundary-layer regulation for oil evaporation is much higher than that for water. These two factors were highlighted two ways:

1) A comparison of the maximum rates of evaporation for some oils, gasoline and water, in the absence of wind, shows that some oil rates exceed that for water by as much as an order of magnitude (water = 0.034 g/min, ASMB = 0.075 g/min, and Gasoline = 0.34 g/min; all under the specific conditions noted), and 2) The saturation concentration of several hydrocarbons in air reveals that some hydrocarbon saturation concentrations in air can be greater than that of water by as much as two orders-of-magnitude.

The fact that oil evaporation is not air boundary-layer regulated implies a simplistic evaporation equation will suffice to describe the process. The following factors do not require consideration: wind velocity, turbulence level, area, and scale size. The factors important to evaporation include time and temperature. Thickness is a factor above certain thicknesses, which are probably not relevant to a rapidly spreading oil slick. The latter is the subject of further experimentation.

REFERENCES

1.   M. Fingas, "Oil and Petroleum Evaporation," Proceedings of the 34th Arctic and Marine Oilspill Program Technical Seminar, Vancouver, 4-6 October 2011, pp. 426- 459.

2.   M. Fingas, "A Literature Review of the Physics and Predictive Modelling of Oil Spill Evaporation," Journal of Hazardous Materials, Vol. 42, No. 2, 1995, pp. 157-175.doi:10.1016/0304-3894(95)00013-K

3.   W. Brutsaert, "Evaporation into the Atmosphere," Reidel Publishing Company, Dordrecht, 1982.

4.   F. E. Jones, "Evaporation of Water," Lewis Publishers, Chelsea, 1992.

5.   J. L. Monteith and M. H. Unsworth, "Principles of Environmental Physics," Hodder and Stoughton, London, 2008.

6.   O. G. Sutton, "Wind Structure and Evaporation in a Turbulent Atmosphere," Proceedings of the Royal Society of London, Vol. 146, No. 858, 1934, pp. 701-722.

7.   D. Mackay and R. S. Matsugu, "Evaporation Rates of Liquid Hydrocarbon Spills on Land and Water," The Canadian Journal

of Chemical Engineering, Vol. 51, No. 4, 1973, pp. 434-439. doi:10.1002/cjce.5450510407

8.   W. Stiver and D. Mackay, "Evaporation Rate of Spills of Hydrocarbons and Petroleum Mixtures," Environmental Science and Technology, Vol. 18, No. 11, 1984, pp. 834- 840.doi:10.1021/es00129a006

9.   Environment Canada, "Online Catalogue of Crude Oil and Oil Product Properties," 2011. http://www.etc-cte.ec.gc.ca/databases/OilProperties/oil_prop_e.html

10.   Z. Wang and M. Fingas, "Oil and Petroleum Product Fingerprinting Analysis by Gas Chromatographic Techniques," In: L. M. L. Nollet, Ed., Chromatographic Analysis of the Environment, Taylor and Francis, Boca Raton, 2005, pp. 1027-1101.

11.   "Ullmann's Encyclopedia," Ullmann Publishing, Hamburg, 2005-2009.

Chapter 8

Experimental Investigation of Fuel Quality and Contaminant Materials from Liquefied Petroleum Fuel

Jae-Kon Kim, Kyong-Il Min, and Eui Soon Yim

Research Institute of Petroleum Technology, Korea Petroleum Quality Distribution & Authority (K-Petro), Cheongju-City, Chungcheongbuk-do, Republic of Korea

ABSTRACT

In this study, the quality characteristics and residue analysis in circulated LPG fuel were investigated experimentally in Korea. Quality characteristics in circulated LPG fuel were examined with samples of LPG in the supply chain (refinery, petrochemical, and imported LPG), transport, gas stations, and vehicles. The experimental results showed that quality of all circulated LPG was well within the quality standard guideline of LPG in Korea. Especially, it has shown average 13 wt ppm in sulfur content over the full circulated LPG. The residue samples in

LPG fuel were extracted on 2 L scale with acetonitrile and analyzed by gas chromatography-mass spectrometry (GC-MS). The components of residues in LPG were composed of 62 organic chemicals with $C_3 \sim C_{28}$ and the main ingredients of residue were plasticizers ((di-octyl phalate (DOP), di-octyl adiphate (DOA) etc.), lubricant oil and amine compounds. It was also showed that mass of residue in vehicles was increasing compared with supply (refinery, petrochemical, and imported LPG). It was presumed that this residue had been originated from automotive LPG fuel, vehicle components, and lubricant oil in infrastructure.

INTRODUCTION

Liquefied petroleum gas (LPG) is another alternative fuel with a high octane number. Alternative automotive fuels with lower carbon content than gasoline and diesel fuel, such as LPG with a mixture of propane (C_3H_8) and butane (C_4H_{10}) [1] , have been widely investigated for reducing exhaust emissions [2] -[8] . Because LPG fuel primarily consists of simple hydrocarbon compounds, and the emissions from LPG-driven vehicles contain lower levels of hydrocarbon compounds, nitrogen oxides, sulfur oxides, air toxins and particulates. LPG is a byproduct generated in petroleum refineries, which is widely used as fuels in residential, industrial and vehicle application. The composition of LPG depends on its end use and varies greatly according to season, country, property of the crude oil/gas supply used and refining process.

Usually, LPG contains certain amount of residues such as hydrocarbons with higher vaporization points falling in the range of lubricants oils ($>C_{18}$) [9] . The residues amount in LPG is an important quality control item and it can accumulate in pipes, vaporizers, instruments and regulators, resulting in mal-flow and misreading. Recently, on specific applications, residue concentrations in LPG have to meet industrial codes. For instance, the Australian LPG Association requires the residues concentration below 20 mass ppm for vehicle use [10].

As a low carbon and low polluting fossil fuel, it is recognized by governments around the world for the contribution, and it can make towards improved indoor and outdoor air quality and reduced greenhouse gas emission [11]. Especially, it is recognized as a clean fuel in Korea and a part of the Korean Government energy supply plan. For this reason, while demand for propane is stagnant, demand for butane has increased since 2000. LPG consumption is 8.63 million tonnes in 2011 and nearly 49% of the LPG consumption is for LPG vehicles. In 2011, 2.43 millions of LPG vehicles were registered to be used throughout the nation, with more than 1948 of LPG gas stations in operation in Korea [12].

The technology for LPG vehicles and LPG fuel in Korea was recognized as a global top. Nevertheless, it recently reported complaints about problems being experienced with the performance of vehicles operating on LPG in Korea. Recently, the problem was being attributed by LPG installer and component suppliers to contaminants in the LPG, an oily substance which was being deposited in vaporizer and injector in vehicles as shown in Figure 1 [13].

Of particular significance was the concern being expressed by a major LPG companies. Under these circumstances, this study was carried out to investigate characteristics of residues in circulated LPG fuel to determine the origin of the offending oily material deposited in the vehicle components, such as vehicle vaporizer and injector in LPG vehicle fuel system.

In this study, the quality characteristics and residue analysis in circulated LPG fuel was investigated experimentally. Quality characteristics in circulated LPG fuel were examined with samples of distribution supply step (refinery, petrochemical, and imported LPG), transport step, gas station step, and vehicle step) in Korea. The residue samples in circulated LPG were extracted with acetonitrile after evaporated on 2 L scale and it was analyzed to determine the origin of residue contaminations by gas chromatography-mass spectrometry (GC-MS) for component analysis.

EXPERIMENTAL

Sample Preparation

The samples of LPG used in the work were obtained from the supply chain as the supply step (refinery, import, petrochemical), transport step, gas station step and vehicle step as shown inFigure 2. The LPG supply system commences at the point where LPG is produced and ends at the point where LPG is dispensed into a vehicle at a gas station. The quality characteristics were tested with 142 samples of LPG in the supply chain for quality specification. The residue component was analyzed with LPG of 2 L to obtain enough residue concentration and in this method sample quantity was increased by 20 times compared with 100 mL as current test method in ASTM D2158-02. The 142 LPG samples for residue analysis were poured in the separatory funnel with 2 L scale and these were evaporated in hood at room temperature for 6 hours. Finally, the residue in separatory funnel was extracted with acetonitrile sometimes and dried to obtain brown oily residue at room temperature as shown inFigure 3.

(a)

(b)

Figure 1: The contamination in vaporizer and injector from LPG vehicles. (a) Vaporizer contamination; (b) Injector contamination.

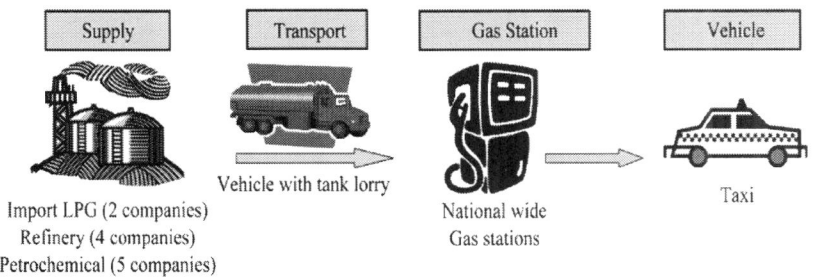

Figure 2: The scheme of experimental sampling of LPG fuel in distribution.

Test Method for LPG Quality

The detailed specification of LPG fuel in this study was listed in Table 1 [14]. The quality of LPG fuel was determined by their physical and fuel properties including composition, sulfur content, vapor pressure, density, residues, and copper corrosion.

LPG Composition

The C3 and C4 composition including olefins in LPG samples were determined by following the KS M ISO 7941, using gas chromatography (GC) (Agilent Technologies 7890A, Palo Alto, USA) equipped with a flame ionization detector (FID) and a chromatography data management system. The fused silica capillary column (SP-2380 model, Supelco, Inc., USA) used in the GC analyzer was 30 m in length and 0.25 mm in inside diameter, with a film thickness of 0.2 μm.

Sulfur Content and Residue Content

For sulfur content, the test was measured by elemental analyzer system NSX-2100V (Mitsubishi Chemical Corporation, Japan) according to ASTM D 6667. The residue was measured by cone-shaped centrifuge tube (Koehler, USA) according to ASTM D 2158 on 100 mL scale.

Vapor Pressure, Density and Copper Corrosion

The vapor pressure was measured by a rapid equilibrium closed cup method (Lawler, USA) according to KS M ISO 4256. The density was measured at 15°C by digital density meter K-25990 (Koehler, USA) using KS M 2150. The copper corrosion was evaluated according to KS M 6151.

All experiments were carried out in triplicate, showing no statistically significant difference between measures. The arithmetic mean of the three determinations was taken as the final result. In addition, the uncertainties were calculated to confirm that each instrument meets the limits of accuracy as set by the specifications of the standard method.

Test Method for LPG Residue Analysis

The condition of GC-MS is showed in Table2 The GC-MS system consisted of an Agilent 7890A GC (Agilent Technologies, Shanghai, China), equipped with an Agilent 7683 auto injector (Agilent Technologies, Shanghai, China), coupled to an Agilent 5975C mass-selective detector (Agilent Technologies, Santa Clara, CA, USA). The GC was fitted with HP-5MS fused silica capillary column (5% phenyl methyl siloxane as non-polar stationary phase, 30 m × 0.25 mm i.d. and 0.25 μm film thickness) from Agilent (J&W Scientific, Folson, CA, USA). The oven temperature program was held at 40°C for 3 min, ramped to 240°C/min and held 240°C for 10 min. Injector and transfer line temperatures were 150°C and 280°C, respectively. The flow rate of the carrier gas (He,99.9995%)was1.5mL/min.Asplitinjectionwithar atioof1:25wasused.Theelectronimpactionizationmassspectrometerw asoperatedasfollows:ionizationvoltage,70eV;ionsourcetemperature,2 00°C;scanmode,50.0-500.0(massrange);scanspeed,1.31scan/sec.The volatileorganic compounds in LPG residue were identified by linear retention indices of a series of C_3 to C_{30} by comparison of themass spectra of each component with the Wiley (Wiley,NewYork,NY,USA) mass spectral library.

(a)

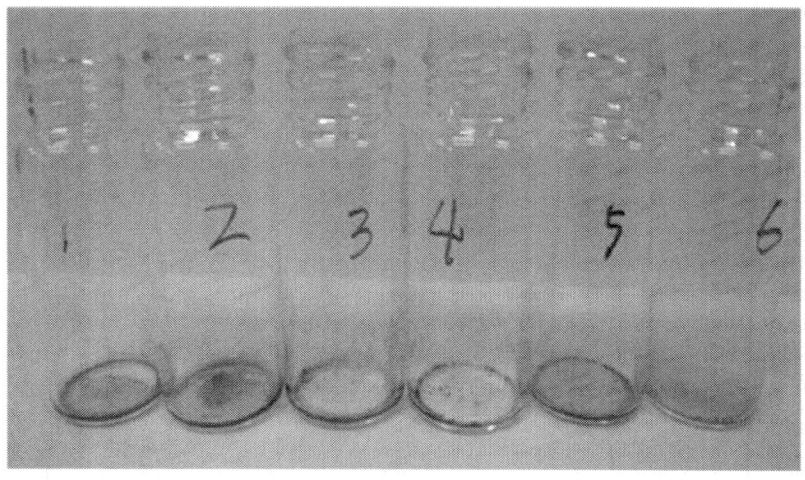

(b)

Figure 3: Extracted residue samples from LPG fuel for GC-MS analysis. (a) Residue in separatory funnel after evaporation; (b) Residue for GC-MS analysis.

Table 1: Testmethodsforqualityspecificationsofautomotive LPG fuelinthisexperimental

Item		LPG standard		Test method
Summer		Winter		
Composition (mol %)	C3-Hydrocarbon	Max. 10	15 ~ 35	
	C4-Hydrocarbon	Min. 85	Min. 60	KS M ISO 7941
	1,3-Butadiene	Max. 0.5		
Sulfur content (mg/kg)		Max. 40		ASTM D 6667
Vapor pressure (40°C, kPa)		Max. 1.27		KS M ISO 4256
Density (15°C, kg/m³)		500 ~ 620		KS M 2150
Residue (ml)		Max. 0.05		ASTM D 2158

| Copper strip corrosion (40°C, 1 h) | Min. 1 | KS M 6251 |

Table2: The conditions of GC-MSf or LPG residue analysis

Item	Condition
GC-MS model	Agilent 7890A GC/EI/MSD (Agilent 5975 C)
Column	HP 5 MS (5% Phenyl methyl siloxane 30 m, 250 µm)
Detector temperature	280°C
Oven temperature condition	40°C (5 min), 5°C/min, 240°C (10 min)
Carrier gas	He 9 kPa, 1.5 mL/min
Injection amount	1.0 µl

Table3: Average results for quality characteristics of automotive LPG samples in distribution

Class Item		LPG standard		Distribution				
Summer		Winter	Supply	Transport	Gas station	Vehicle	Average	
Composition (mol %)	C3-Hydrocarbon	Max. 10	15 ~ 35	1.1	10.3	14.1	6.8	8.7
	C4-Hydrocarbon	Min.85	Min. 60	98.4	91.2	85.3	92.1	91.1
	1,3-Butadiene	Max. 0.5		0.04	-	-	0.06	0.02
Sulfur content (mg/kg)		Max. 40		8.1	14.7	16.4	10.0	12.9
Vapor pressure ($40°C$, kPa)		Max. 1.27		0.45	0.58	0.57	0.42	0.52
Density ($15°C$, kg/m^3)		500 ~ 620		578	570	569	574	572
Residue (ml)		Max. 0.05		0.05↓	Max. 0.05	Max. 0.05	Max. 0.05	Max. 0.05
Copper strip corrosion ($40°C$, 1 h)		1		Max. 1	Max. 1	Max. 1	Max. 1	Max. 1

RESULT AND DISCUSSION

Quality Characteristics of LPG in Distribution

The quality characteristics of LPG fuel were tested with 142 samples in distribution. The experimental results showed that quality of all circulated LPG fuel was well within the limit by Korean specifications as shown in Table 3 Figure 4 shows a general GC chromatogram for composition analysis of butane in LPG fuel. The LPG fuel composition was evaluated 8.7 mol% for C_3 hydrocarbon and 91.1 mol% for C_4 hydrocarbon by GC. Especially, the residue of LPG fuel was within the limit by Korean specifications on 100 mL scale as a current test method (ASTM D 2158) and was showed average 13 wt ppm in sulfur content over the circulated LPG fuel.

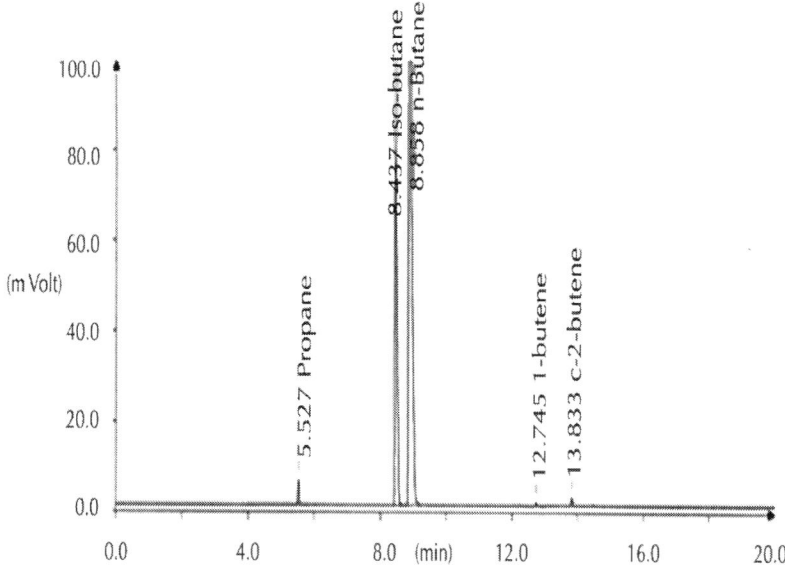

(a)

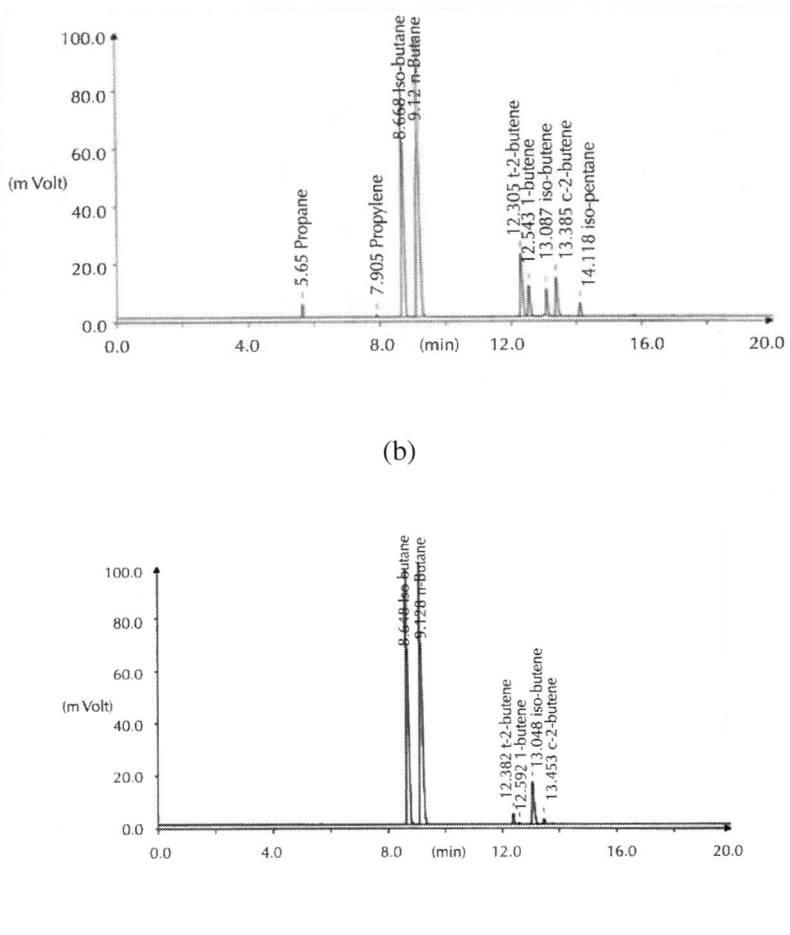

(b)

(c)

Figure 4: GC chromatogram for butane composition analysis of LPG fuel in supply step. (a) Importing butane; (b) Refinery butane; (c) Petrochemical butane.

Figure 5 shows a general GC chromatogram for olefin analysis of LPG in vehicle step. The olefin means a sum of unsaturated hydrocarbon such as t-2-butene, 1-butene, iso-butylene, c-2-butene, iso-butene, 1, 3-butadiene in LPG composition. Kim et al reported that the olefin content with unsaturated hydrocarbons in LPG has an effect to extract plasticizers [15] [16] . The total olefin composition in distribution also was analyzed to be much as 3.2 mol% average but it was a high level as 7.31 mol% in vehicle step compared with other step as shown in Figure 6. It was statistically increased in the order of transport step < gas station step < vehicle step compared to supply step in distribution.

Residues Analysis of LPG in Distribution

The sources of the residues are the LPG processing equipment, compressors and containers. The amount of residue in LPG is an important quality control item, since it can accumulate in pipes, vaporizers, injectors, instruments and regulators, resulting in mal-flow and misreading. In Korea, the residues content in LPG is measured by ASTM D2158-02 test method. However, it can't give an enough residues concentration to analyze residue component because of a small amount of 100 mL in sampling. To investigate the characteristics of LPG residues for concentration, LPG fuel was evaporated with LPG of 2 L scale to get a more sample compared with contaminants causing problems experienced with the performance of vehicles operating. The residue analysis was tested with 142 samples of LPG in distribution.

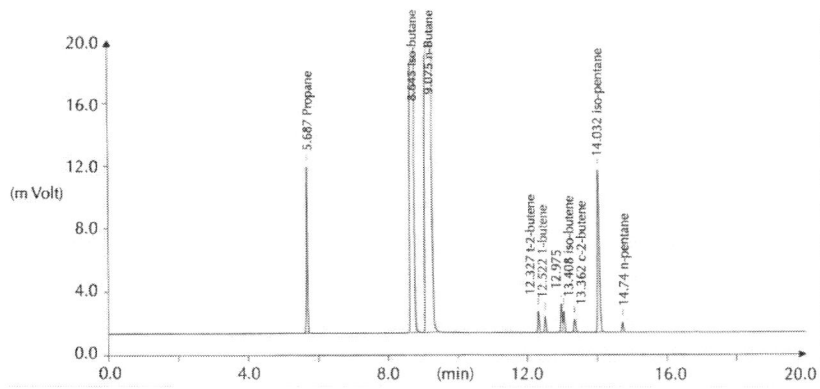

Figure 5: GC chromatogram for butane composition analysis of LPG fuel in vehicle step.

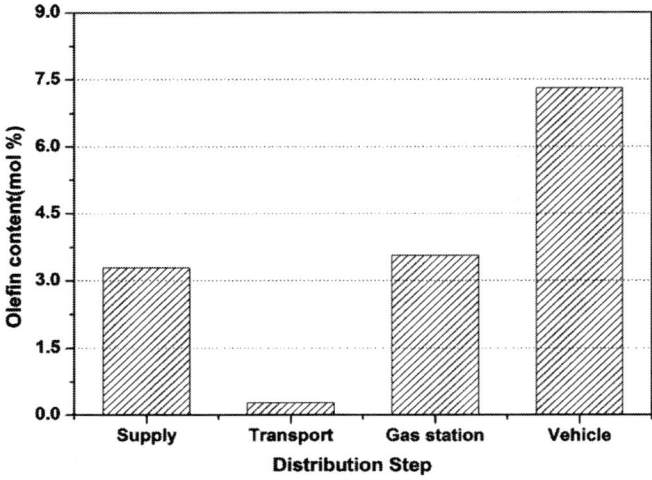

Figure 6: Total olefin content comparison of LPG (142 samples) in distribution (the olefin means a sum of unsaturated hydrocarbon in LPG composition).

The amount of residues comparison of LPG on 2 L scale in distribution is shown in Figure 7. The average total amount of residues in distribution is 19.58 mg/2L and the amount of residue in vehicle step (45.63 mg/2L) is higher than supply step (16.91 mg/2L). It showed that the amount of residue is increasing due to contamination on supply in distribution. Especially, Figure 8indicated that the residue content of imported LPG was lower than refinery LPG, petrochemical LPG in supply step. The components of residues in LPG were detected62 hydrocarbon chemicals with $C_3 \sim C_{28}$ same as plasticizers, alkane, alkenes and ester with high molecular formula in lubricant oil, amine components, sulfides and siloxane in Table4 Especially, the main ingredients of residue were analyzed plasticizers same as di-n-octylphalate, di-octyladiphate, di (2-ethylhexyl)adipate, di-isooctylalipate, di(isodecyl)phthalate, di-heptyl phthalate by GC-MS in Table5

In general, plasticizers are used in the manufacture of plastics and rubbers to facilitate manufacture and make the products more flexible and softer to meet specific application requirements. Most plasticizers are liquid in their natural state and have an appearance similar to lubricant oil. The harmful plasticizer comprises mostly of either phthalates or adipates or a combination of these chemicals and they were found to have a detrimental effect on the rubber based components. Therefore, LPG residues with high molecular weight have an effect on fuel distribution system in Korea.

CONCLUSIONS

The objective of this research was to assess an investigation of characteristic of residues in circulated LPG fuel

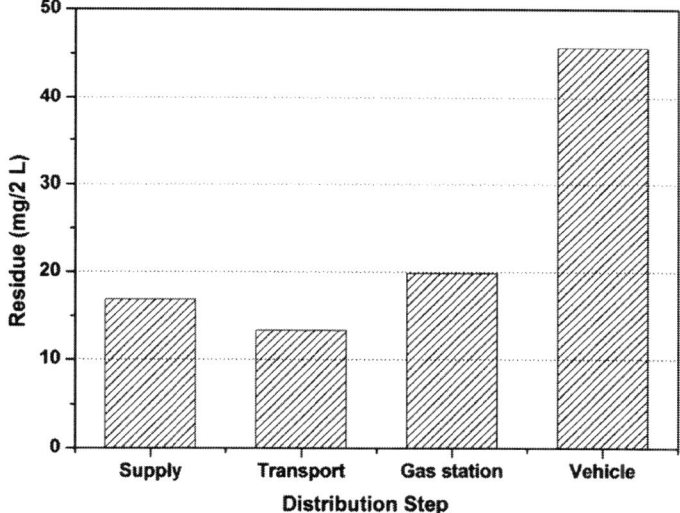

Figure 7: Total amount of residues (mg/2L) comparison of LPG (142 samples) in distribution.

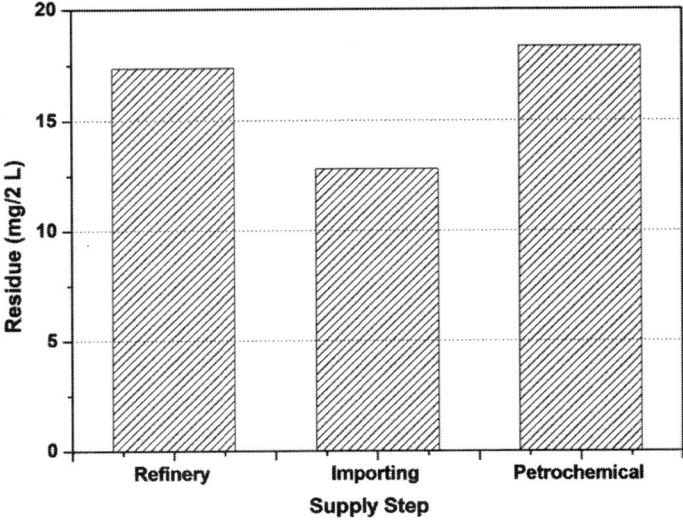

Figure 8: The amount of residues (mg/2L) comparison of LPG (142 samples) in supply step.

Table 4: The residue components from automotive LPG fuel in distribution

Class	Component	Molecular formula	Detection step in distribution[a]
Alkanes	2,2-dimethyl butane	C_6H_{14}	G
	Isodecane	$C_{10}H_{22}$	G
	Undecane	$C_{11}H_{24}$	S
	2,2,6,6-tetramethylheptane	$C_{11}H_{24}$	G
	Dodecane	$C_{12}H_{26}$	G
	Tridecane	$C_{13}H_{28}$	G
	2,6,7-trimethyl-decane	$C_{13}H_{28}$	G
	Hexadecane	$C_{16}H_{34}$	G
	Tetracosane	$C_{24}H_{50}$	G
	n-Hexacosane	$C_{26}H_{54}$	G
	n-Octacosane	$C_{28}H_{58}$	S

Amines	Ethamine	C_2H_7N	S
	1,3,4-Thiadiazol-2-amine	$C_2H_3N_3S$	S
	1,2-Ethanediamine	$C_2H_8N_2$	G
	Silanamine (H_5NSi)	H_5NS	G
	N,N-dimethyl formamide	C_3H_7NO	G
	Methyldiethanol amine (MDEA)	$C_5H_{13}NO_2$	G
	2,4,6-Trimehtyl-5H-1,3,5-dithiazin	$C_6H_{13}NS_2$	S
	2-heptanamine	$C_7H_{17}N$	S
	Diphenylamine	$C_{12}H_{11}N$	T
	Octyl-diphenylamine	$C_{20}H_{27}N$	T
Alkens	2,4,6-dodecatriene	$C_{12}H_{20}$	S
	Dimethyl-formyl thiacyclohexene	$C_8H_{13}SO$	G
Siloxane	Ttrisiloxane	$H_8O_2Si_3$	S
	Pentasiloxane	$H_{12}O_4Si_5$	G
	Cyclohexasiloxane	$H_{12}O_6Si_6$	S, G
Sulfides	Diethyldisulfide	$C_4H_{10}S_2$	T
	Methyl tert-butyl disulfide	$C_5H_{12}S_2$	T
	Bis(1,1-dimethylethyl) disulfide	$C_8H_{18}S_2$	T
	Dibutyldisulfide	$C_8H_{18}S_2$	S
	Di-n-octyl disulfide	$C_{16}H_{34}S_2$	S
Esters	Methyl cis-3-chloropropenate	$C_4H_5ClO_2$	G
	Hexadecanoic acid methyl ester	$C_{17}H_{34}O_2$	G
	Hexadecanoic acid methyl ester	$C_{19}H_{36}O_2$	G
	Hexanedioic acid dioctyl ester	$C_{22}H_{42}O_4$	S

Table 5: The main ingredients of residues contaminations from automotive LPG fuel in distribution

Component	Molecular formula	Detection step in distribution[a]
Di(2-ethylhexyl) adipate	$C_{22}H_{40}O_4$	S, T, G, V
Di-isooctylalipate	$C_{22}H_{42}O_4$	S, V
Di(isodecyl)phthalate	$C_{28}H_{44}O_4$	G
Di-heptyl phthalate	$C_{22}H_{34}O_4$	G
Bis(n-octyl)phthalate	$C_{24}H_{36}O_4$	G, T
Di(2-ethylhexyl) phthalate	$C_{24}H_{38}O_4$	S, T, G
Di-nonyl phthalate	$C_{26}H_{42}O_4$	G
Di(7-methyloctyl) phthalate	$C_{26}H_{40}O_4$	G

[a]S: supply step, T: transport step, G: gas station step, V: vehicle step.

[a]S: Supply step, T: Transport step, G: Gas station step, V: Vehicle step.

to determine the origin of the offending oily material which was being deposited in the vehicle components such as vehicle vaporizer and injector in vehicle fuel system. Recently, the problem was being attributed by LPG installer and component suppliers to contaminants in the LPG an oily substance which deposited in vaporizer and injector in vehicle.

The experimental results showed that quality of all circulated LPG was well within quality standard guideline of LPG in Korea. Especially, it showed average 13 wt ppm in sulfur content over the full circulated LPG. The components of residues in LPG was detected 62 organic chemicals with $C_3 \sim C_{28}$ hydrocarbon compounds and the main ingredients of residue were plasticizers ((di-octylphalate (DOP), di-octyladiphate (DOA) etc.), lubricant oil and amine components. Finally, it was presumed that this residue had been originated from automotive LPG fuel, vehicle components and lubricant oil in infrastructure. LPG residue with high molecular weight has an effect on fuel distribution system in Korea.

ACKNOWLEDGEMENTS

This research was supported by Korea LPG Association (KLPGA) and we sincerely appreciate the support of KLPGA.

REFERENCES

1.   Keith, O.K. (1995) Automotive Fuels Reference Book. Society of Automotive Engineers, USA.

2.   Boretti, A. (2009) Development of a Direct Injection High Efficiency Liquid Phase LPG Spark Ignition Engine. SAE Paper, No. 2009-01-1881.

3.   Chiriac, R., Raud, R., Niculescu, D. and Apostolescu, N. (2003) Development of a LPG Fueled Engine for Heavy Duty Vhicles. SAE Paper, No. 2003-01-3261.

4.   Goto, S., Lee, D., Shakal, K., Harayama, N. and Ueno, H. (1999) Performance and Emission of an LPG Lean-Burn Engine for Heavy Duty Vehicles. SAE Paper, No. 1999-01-1513.

5.   Kim, C., Lee, D., Oh, S., Kang, K., Chol, H. and Min, K. (2002) Enhancing Performance and Combustion of an LPG MPI Engine for Heavy Duty Vehicles. SAE Paper, No. 2002-01-0449.

6.   Lee, J.W., Do, H.S., Kweon, S.I., Park, K.K. and Hong, J.H. (2010) Effect of Various LPG Supply Systems on Exhaust Particle. International Journal of Automotive Technology, 6, 119-124.

7.   Myung, C.L., Lee, H., Choi, K., Lee, Y.J. and Park, S. (2009) Effects of Gasoline, Diesel, LPG and Low-Carbon Fuels and Various Certification Modes on Nanoparticle Emission Characteristics in Light-Duty Vehicles. International Journal of Automotive Technology, 10, 537-544. http://dx.doi.org/10.1007/s12239-009-0062-9

8.   Park, K. and Han, S. (2004) The Ignition Characteristics for the Direct Liquid Injection Combustion in LPG Engine. FISITA Paper, No. 20042357.

9.   Zhuang, Q., Yodatani, J. and Kato, M. (2005) Accurate Measurement Method for the Residues in Liquefied Petroleum

Gas (LPG). Fuel, 84, 443-446.http://dx.doi.org/10.1016/j.fuel.2004.09.005

10. Government of Western Australia (2008) Investigation of the Cause and Consequences of Contaminants in Autogas Vehicle Systems. 2008 Report.

11. Goto, S., Wakasa, R. and Lee, D. (2001) Research Trends of LPG-Fueled Engine System. Journal of the Society of Automotive Engineers of Japan, 55, 30-37.

12. Korea LPG Association (2011) Statistical Review of Global LP Gas. 2011-Report.

13. Korea LPG Association (2009) Study of the Origin of Unknown Materials in LPG Distribution. 2009 Report.

14. Korea Petroleum Quality & Distribution Authority (2014) Fuel Quality Standards for Liquefied Petroleum Gas in Korea. http://www.kpetro.or.kr/sub.jsp?MenuID=m2as402

15. Kim, Y.U. (2002) The Component and Compositional Analysis of Trace Materials in LPG. Journal of the Korean Chemical Society, 46, 317-322.http://dx.doi.org/10.5012/jkcs.2002.46.4.317

16. Kim Y.U. (2004) Extraction Property of Plasticizer in LPG High Pressure Rubber Hose. Journal of the Korean Chemical Society, 48, 156-160.http://dx.doi.org/10.5012/jkcs.2004.48.2.156

Chapter 9

Environmental Consequences of Long-Term Development of Petroleum Fields, Absheron P-la, Azerbaijan, Case History

Akper A. Feyzullayev and Vagif B. Ibragimov

Geology Institute of Azerbaijan National Academy of Sciences, Baku, Azerbaijan

ABSTRACT

In this paper the overview of a level of study of the deformation processes on long-term petroleum field development and ecological consequences accompanying them is given. In more details these processes are considered for the oldest Absheron petroleum bearing region (Absheron p-la, Azerbaijan) where fields are strongly depleted and the pressure drop reaches 80% from initial values. Ecological consequences of this phenomenon are the following: development in area of petroleum fields Balakhany-Sabunchi-Ramany, Surakhany, Garachukhur, Bibi-Eybat intensive pro- cess of ground subsiding (up to

47 mm/year) and flooding, frequent incidences of curved boreholes, breaks in oil, gas, and water pipelines, and sudden kicks of water and sand, occurrence of the induced seismicity (Surakhany earthquake in 1937 with magnitude 6). With purpose of forecasting of geodynamic processes creation on petroleum production complexes of system of the environmental control is recommended.

INTRODUCTION

As is known large amounts of oil, gas and associate water are extracted from the subsoil when carrying out long term hydrocarbon field development. This results in the significant pressure decline in the reservoirs which, according to different estimates, is 50% - 80% below hydrostatic pressure [1]. Such pressures are classified as abnormally low reservoir pressures (ALRP) [2].

The reservoir pressure decline leads to the increase of the effective pressure which in its turn causes deformation processes such as decrease of rock porosity and permeability as well as additional compaction [3]. This process finds its reflection in the change of the environmental situation in the oil fields which is manifested in the soil subsidence [4] [5], under flooding [6], appearance of "induced" seismicity [7]; that finally can cause damages for the infrastructure (well, platform and pipeline failure and others) and require large investments for its rehabilitation [8].

In this paper an overview of the problem study in the world and analysis of some of its aspects for the Absheron peninsula (Azerbaijan) (Figure 1), where oil and gas fields have been developed for more than a century and are currently considerably depleted, is provided.

OVERVIEW OF PROBLEM

Land Subsidence and Flooding

There are distinguished three types of manmade extraction of subsoil fluids which under favorable geological conditions can cause significant LS: 1) oil, gas and associated water production; 2) hot water or steam extraction for the purpose of using geothermal energy 3) ground water extraction. Each of them causes maximal LS nearly of the same order of magnitude. Thus, for instance, the LS at the well known Wilmington oil field (California State, USA) reached 9 meters. Hot water extraction for the purpose of using geothermal energy in Wairakei (New Zealand) led to 6 - 7 meters of subsidence, and ground water extraction in Mexico (Mexico) and in San Joaquin valley in California (USA) caused the 9 meter LS [9] .

Land subsidence connected with intensive hydrocarbons production is observed in many regions of the world. Such phenomenon was for the first time recorded in 1926 at the Goose Creek oil field not far from Houston (Texas State, USA). Here the oil production from loose sandstones and silts from the depth of 350 - 1400 m (thickness of the oil bearing interval more than 300 m) during the period of 1917-1925 led to 1 m subsidence [10] - [12] . Later, the similar effect was identified at other areas. Particularly, the rate of changeable in space and time LS at Belridge and Lost Hills oil fields in California (USA) reaches 30 - 40 cm annually [13]. In Houston- Galveston (Texas area, USA), where there are at least 110 oil and gas fields, 29 fields have been crossed by the lines of relevelling. Six of them have been discovered to have local land subsidence: Alco-Mag, Chocolate Bayou, Goose Creek, Hastings, Mykawa and South Houston, whereas the LS rates (up to 120 mm/year) exceeded the natural subsidence rates estimated to be 13 mm/year [12]. It is thought that such rapid LS rates have been caused by the large scale removal of fluids forming a large subsidence cup [14].

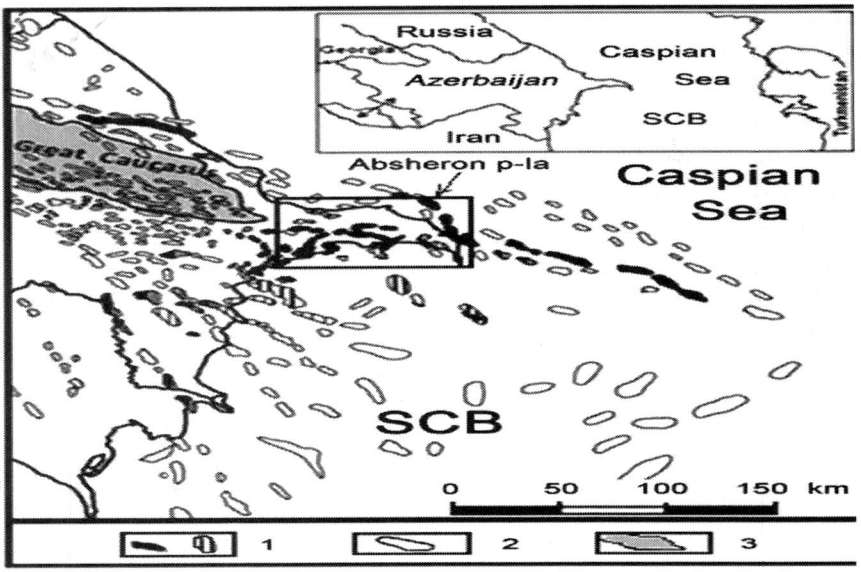

Figure 1: Location of Absheron p-la, Azerbaijan. 1: petroleum fields; 2: structures; 3: mountains.

This phenomenon is thought to have two main reasons: **a)** natural, related with geological process of sediments compaction, and **b)** manmade ones. The first ones though widely spread are characterized by relatively low rate of LS compared with the second ones. For instance, the study of the history of development and reservoir pressure at some large oil and gas fields in Louisiana state (USA) suggested that the rates of the man-in- duced LS reach 9 mm/year or 23 mm/year (depending on whether these data have been obtained based on the results of releveling or sea level records), whereas geological rates of LS come up to about 3 mm/year [4] [15]. The rates of LS at the Port Neches field in Texas (USA) in the period of maximal petroleum production and discovered that the computed rate of land subsidence (30 mm/year) are 3 order higher than the geological rates of LS at the coastal valley [15]. The average historical rate of LS in the Mississippi delta is 12mm/year [16] which is higher than the average geological rates of LS [17].

Considerable LS at the developed hydrocarbon pools is observed under the presence of all or some of the following conditions: 1) essential drop of the reservoir pressure during the hydrocarbons production

[22]; 2) hydrocarbons production from wide depth interval [18]; 3) oil or gas (separately or together) are contained in slightly consolidated or slightly cemented rocks; 4) reservoirs have a relatively shallow depth occurrence and considerable extension.

Collection, classification and analysis of the published data on more than 130 hydrocarbons pools under production have discovered the LS occurrence from tens of centimeters to several meters [19] [20]. Special studies have established the following characteristic features of the considered phenomenon:

-The rapid decline of the hydrocarbons production leads to the reduction of LS rates. So, as a rule, the initial intensive period of the field development is characterized by higher LS rates which later slow down whereas in some cases subsidence stops.

-LS on the developed oil and gas fields has usually local character. However, hydrocarbon production from one bed on several relatively near fields can lead to the regional pressure loss and regional LS [21].

-LS can be restrained by means of reservoir pressure recovery/maintenance. At the same time, the process of recovery can be successful if only all fault blocks work as a whole. According to some researchers actions aimed at maintaining the reservoir pressure by means of water injection into the bed and other methods applied at many fields worldwide allow to recovery pressure not exceeding 10% of initial value. It is connected with the fact that inelastic deformations of rocks accompanying the reservoir pressure decline are more significant than the elastic ones which leads to irreversible process of reservoir rocks consolidation [3]. Therefore restore the reservoir pressure to the initial value will not lead to the recovery of the initial reservoir capacity [22].

Widely spread processes of reservoir compaction and subsidence belong to the categories of dangerous natural phenomena which are inevitably accompanied by serious environmental and economic consequences. The economic consequences include flooded territories due to LS especially in coastal areas where even slight LS can cause flooding. Such phenomenon was for the first time described for the oil field near Houston (Texas State, USA). The land subsidence which reached almost 9 meters due to the oil extraction from the main Wilmington field [9] led to the flooding of streets and piers in the city and in the Long Beach port (California, USA), caused damages to the bridges, railways and other port facilities [23] . The smaller LS (~1

m) at the Goose Creek field (Texas, USA) was enough to change the landscape from the green highland to open water. LS in Venezuela led to the huge flood at the coastal area of the Maracaibo Lake, and in Russia large oil bearing areas in Western Siberia turned into swamps due to similar reasons [6].

Economic consequences of LS as the result of hydrocarbons production often cause serious damages to the existing infrastructure (well, pipeline and platform failure etc.). According to one of the present day estimates the annual damage caused by LS due to hydrocarbons production within the USA comes up to more than 100 million dollars [8]. Here should also be added the so called "postponed" damage which will be quite appreciable by future generations: it has been established that 2 - 3 m of LS lead to the reduction of crop yield by 10%, 5 - 6 m—by 50%, and more than 8 m—cropland is totally destroyed [24] .

Induced Seismicity

It has been established that man's activity can also provoke emergence or intensification of seismic activity in the areas of hydrocarbons production. Such seismicity was named as man caused. In literature it is often referred to as "caused", "induced", "excited", "generated" etc. With the purpose of unification of terminology Russian scientists V. V. Adushkin and S. B. Turuntaev [7] offers all seismic events which occur due to various reasons after or in the process of man caused impact to define as induced seismicity.

The occurrence and intensity of induced seismicity at developed oil and gas fields depend on the geological- tectonic properties of the reservoir and surrounding massif, its mode of deformation, degree of blockiness and presence of heterogeneity, rates and volumes of hydrocarbons extraction and water injection [25]. According to the analysis of 200 fields situated in different regions of the world the likelihood of induced subsidence occurrence increases with the increase of depth and thickness of the developed bed as well as with the decline of reservoir porosity and permeability. Therewith, focuses of induced seismicity occur at various depths (depending on the geological structure of the massif) over and below the productive zones, and its strengthening happens in case of misbalance between the volume of extracted oil and injected liquid.

Researches show, that the interval of time between the beginning of development of petroleum fields and the provoked activization of seismic activity for gas fields is less (2 - 16 years), than for oil (7 - 39 years) [26].

The examples of induced seismicity are well known: it are disastrous earthquakes in the area of the Gazli gas field in 1976 and in 1984 with magnitudes from 6.8 to 7.3, and the Neftegorsk earthquake in 1995 with the magnitude of 7.2 - 7.6 which was the consequence of the active oil production on the Sakhalin island [27] .

RESULTS AND DISCUSSION

As is known the history of oil production on the territory of the Absheron peninsula (Azerbaijan) goes back to the middle of the XIX century when oil was extracted from shallow wells dug by hand. Later with the introduction of well drilling technology the intensity of hydrocarbons extraction from reservoirs was continuously increasing and by the end of the XX century exceeded one billion tons of oil (plus the extraction of large amounts of associated gas and reservoir water). This resulted in the abrupt decline of reservoir pressures.

The analysis suggests that in operated sites of the productive stratum of the Absheron peninsula the initial reservoir pressures which approximately corresponded to hydrostatic ones dropped from 52.3% (Gala field) to 80.6% (Chakhnaglar-Sulutepe field). Whereas the level of primary pressure drop in the reservoirs of the upper formations of the productive stratum is higher (in average about 66%) than in the reservoirs of the lower formations (58.9%) of the productive stratum.

The level of present day pressures in comparison with the primary ones is shown on the diagram of their change with depth built according to the data of about 15 fields of Absheron peninsula (Figure 2).

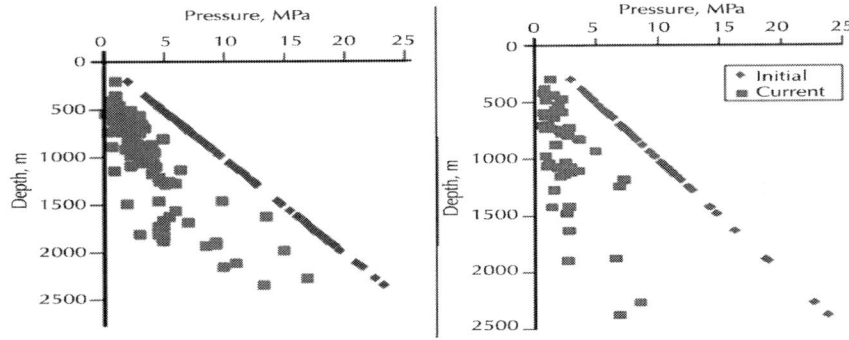

Figure 2: Change with depth of initial and current (as of 01.01.2009) reservoir pressures at the fields of Absheron peninsula (on left) and at oldest and largest Balakhany-Sabunchi-Ramany field, which placed in operation in 1873-1890.

The development of corresponding deformation processes which can be identified by special studies for example monitoring of present day vertical movements of the earth crust must have been the natural consequence of the reservoir pressure decline in the fields of the Absheron peninsula. Such monitoring at the Absheron peninsula was launched as early as in 1912 when the geodynamic polygon being one of the first in the world was established here. The purpose of the first geodetic measurements was to identify differences in terms of altitude of the levels of the Caspian and Black seas.

While processing the relevelling data it was discovered that separate areas of the Absheron peninsula vary with altitude [28]. In particular, attention was paid to the subsidence of the central eastern parts of the peninsular (Figure 3). Later there was detected subsidence of the Bibi-Eybat oil field territory. For instance, LS with the speed of 30 mm/year was detected in the area of one of the oldest wells No. 2465, and in the area of the well No. 2778—31 mm/year. The similar phenomenon was observed at other fields of the peninsular (Table 1). In absolute values the presence of abnormal cases in fold areas of anticlines at the oldest oil fields in Sabunchi, Surakhany, Ramany and Bibi-Eybat reached 1 - 2.5 m for the period of 50 years, and during 80 years the central part of the Absheron peninsular subsided by more than 3 m [29] .

Though the analysis of the results of relevelling carried out on the Absheron peninsula did not contribute to unanimous opinion as to the causes of LS, however, a number of scientists link one of them to the intensive extraction of oil, gas and sand from the subsoil. Thus, according to the observations of D.A. Liliyenberg and others [30] the periods of relative tectonic LS at the fields of Binagadi, Surakhany and Bibi-Eybat correspond to the sudden increase of hydrocarbons extraction from the subsoil, whereas the periods of movement stabilization or relative elevation are associated with minimal production. The diagram shown in Figure 4 illustrates the noticeable tendency for the increase of average annual rate of the vertical earth crust movements accompanied by the increase of the average annual oil production at the fields of the Absheron peninsula (without accounting of extraction of gas and associated reservoir waters).

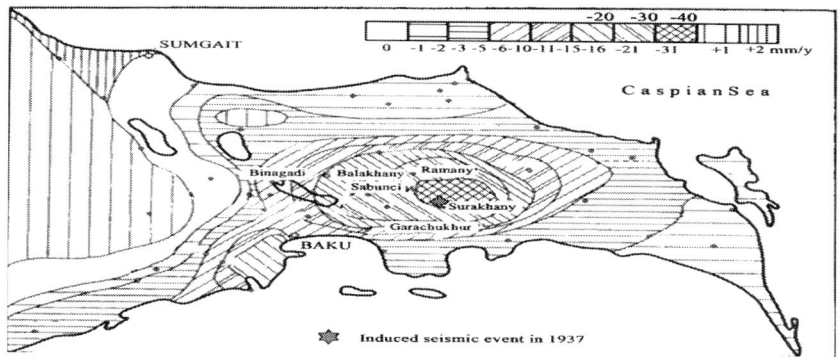

Figure 3: Scheme of vertical movements on Absheron p-la based on results repeated leveling.

Table 1: Land subsidence at some fields of the Absheron peninsular

Field	Years of levelling	Time interval, years	Annual average rate, mm/year
Binagadi	1912-1937	25	-2
Bibi-Eybat	1908-1972	64	-30
Surakhany	1926-1974	48	-30

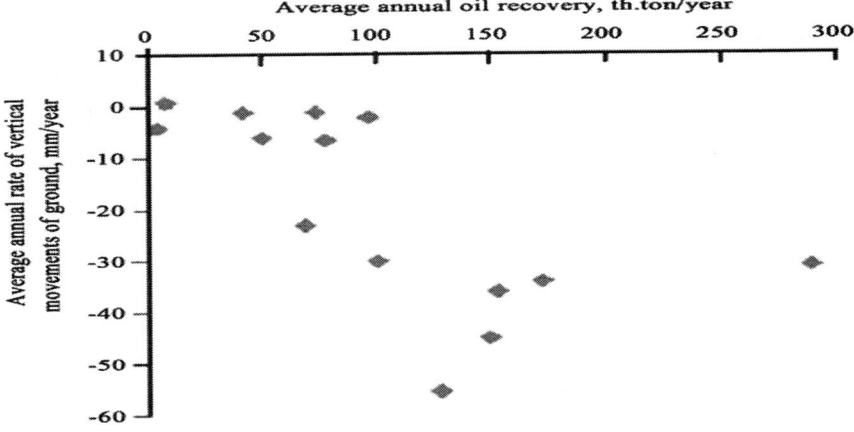

Figure 4: Diagram of dependence of earth crust movement speed on average annual oil production at the fields of Absheron peninsula (created based on Yashenko's data [28]).

The above pointed processes of land subsidence in the Absheron peninsula as well as at long term developed oil fields of other basins in the world were also accompanied by negative environmental consequences. Thus, according to the data of the hydrodynamic monitoring carried out in the Absheron peninsula there have been identified areas of intense rising of level of underground water correlating to LS areas. In particular, in the central part of the peninsular the area of zones with the depth of ground water level of >10 m decreased by 21.4% during the period from 1955 to 2006, and the area of zones with the depth of ground water level of 3 - 5 m increased by 10.3%. This serves as a vivid example of the negative impact of the long term development of oil and gas fields in this territory for hydrogeological and ecological environments which is manifested in flooding (in other cases-swamping) of territories and obvious deterioration of ameliorative properties of lands.

Abnormal rates of LS in the number of fields of the Absheron peninsula can serve as a cause of the man-in- duced seismicity. Abnormal rates of subsiding in area of Surakhany field (about 47 mm/year) have served as the reason for earthquake of magnitude 6 which occurred in 1937 near the same named village (see Figure 3).

Oil and gas production at the intensely developed oil fields of the Absheron peninsula was often accompanied by cases of borehole curving and breaking, oil, gas and water pipeline rupture, sudden water and sand kick etc. which were the results of man induced deformation processes combined with natural geodynamic processes. Thus, for example in the Balakhany-Sabunchi-Ramany field in more than 20 wells there were reported cases of breakage (wells No. 1794, 2304 and others) and breakage of borehole lines (wells No. 2409, 2575, 2610 and others) as well as their curving (wells No. 1499, 12204 and others) which led to serious economic damages [28] .

Well failures are causing large scale contamination of oil-field territories too. Two monitoring conducted in Azerbaijan over the last years by the Ministry of Ecology and Natural Resources identified around 30 thous. ha of contaminated land of which 10 thous. ha (according to other data—12 thous. ha) fall on land contaminated with oil and oil products essentially in the Absheron peninsula.

It is known that on completing the field development oil/gas wells used for the production of hydrocarbons are abandoned or temporary suspended. However, the extent of risk of dangerous (emergency) situations will not decrease as the control over their condition is often either not exercised or exercised quite cursorily without taking into account all the requirements of relevant regulatory acts [31].

According to the independent experts (no official data—Authors) in Azerbaijan only on the territory of the Absheron peninsular there is a concentration of up to 3 thousand abandoned i.e. uncontrolled oil wells.

A more detailed analysis of these and a number of other environmental problems is the subject of separate researches.

CONCLUSIONS

Overview of references has shown that intensive development of oil-and-gas fields causes natural disbalance and wide progress of deformation processes: decline of reservoir pressure induced additional compaction of rocks and as a result, subsidence and displacement of a ground occur.

The above mentioned undesirable environmental consequences of the long term development of oil and gas fields in the aggregate cause considerable economic damage to the existing infrastructure. For minimization of such damage it is necessary to equip petroleum producing complexes by system of environmental control with purpose of forecasting geodynamic processes development.

REFERENCES

1.   Xie, X., Jiao, J.J., Tang, Z. and Zheng, C. (2003) Evolution of Abnormally Low Pressure and Its Implications for the Hydrocarbon System in the Southeast Uplift Zone of Songliao Basin, China. AAPG Bulletin, 1, 99-119.

2.   Koshlyak, V.A. (2002) Granitoid Oil and Gas Reservoirs. Tau-Press, Ufa.

3.   Imanov, A.A. (2012) Physical and Mechanical Properties Sedimentary Rocks of the South-Caspian Basin in Subsurface Conditions: Deep Water Hydrocarbons Resources. Nafta-Press, Baku.

4.   Morton, R.A. and Buster, N.A. (2002) Subsurface Controls on Historical Subsidence Rates and Associated Wetland Loss in South-Central Louisiana. Gulf Coast Association of Geological Societies, 52, 767-778.

5.   Patel, P. and Kulkarni, M.N. (2007) Application of GPS for Monitoring Land Subsidence. Journal of Earth Sciences, 1, 35-46.

6.   Serebryakov, V.A. and Chilingar, G.V. (2000) Prediction of Subsidence: Relationship between Lowering of Formation Pressure and Subsidence Due to Fluid Withdrawal. Energy Sources Part A: Recovery, Utilization and Environmental Effects, 5, 409-416.

7.   Adushkin, V.V. and Turuntaev, S.B. (2005) Man Induced Processes in the Earth Crust (Hazards and Catastrophes). INEK, Moscow.

8.   National Research Council (1991) Mitigating Losses from Land Subsidence in the United States. National Academy Press, Washington DC.

9.   Mayuga, M.N. and Allen, D.R. (1970) Subsidence in the Wilmington Oil Field, Long Beach, California, U.S.A. In: Tison,

L.J., Ed., Land Subsidence, International Association of Scientific Hydrology, UNESCO, 66-79.

10. Pratt, W.E. and Johnson, D.W. (1926) Local Subsidence of the Goose Creek Oil Field. Journal of Geology, 34, 577- 590.http://dx.doi.org/10.1086/623352

11. Snider, L.C. (1927) A Suggested Explanation for the Surface Subsidence in the Goose Creek Oil and Gas Field, Texas. AAPG Bulletin, 11, 729.

12. Holzer, T.L. (1990) Land Subsidence Caused by Withdrawal of Oil and Gas in the Gulf Coastal Plain, the Houston, Texas, Case History. AAPG Bulletin, 74, 1497-1498.

13. Van der Kooij, M. (1997) Land Subsidence Measurements at the Belridge Oil Fields from ERS InSAR Data. The 3rd ESA ERS Symposium, Florence.http://earth.esa.int/workshops/ers97/papers/vanderkooij1/index-2.html

14. Gabrysch, R.K. and Coplin, L.S. (1990) Land-Surface Subsidence in Houston-Galveston Region, Texas. Texas Water Development Board Report, No. 188, Austin.

15. Morton, R.A., Purcell, N.A. and Peterson, R.L. (2001) Field Evidence of Subsidence and Faulting Induced by Hydrocarbon Production in Coastal Southeast Texas. Gulf Coast Association of Geological Societies Transactions, 51, 239- 248.

16. Shinkle, K. and Dokka, R.K. (2004) Rates of Vertical Displacement at Benchmarks in the Lower Mississippi Valley and the Northern Gulf Coast. NOAA Technical Report, No. 50, 135 p.

17. Penland, S., Ramsey, K.E., McBride, R.A., Mestayer, J.T. and Westphal, K.A. (1988) Relative Sea-Level Rise and Delta-Plain Development in the Terrebonne Parish Region. Coastal Geology Tech Report, No. 4, 121 p.

18. Hejmanowski, R. (1993) Zur Vorausberechnung förderbedingter Bodensenkungen über Erdölund Erdgaslagerstätten (Prediction of Land Subsidence Induced by Oil and Gas Exploitation). Ph.D. Thesis, Clausthal University of Technology, Clausthal-Zellerfeld.

19. Melnikov, N.N. and Kalashnik, A.I. (2009) Man Induced Geodynamic Processes When Developing Oil and Gas Fields of the Barents Sea Shelf. Vestnik MGTU, 12, 601-608.

20. Nesterenko, Y.M., Kosolapov, O.V. and Nesterenko, M.Y. (2010) Seismic Activity in the Regions of the Developed Hydrocarbons Fields. News of the Samara Scientific Center of the RAS, 12, 1240-1244.

21. Kreitler, C.W., Akhter, M.S., Donnelly, A.C.A. and Wood, W.T. (1988) Hydrogeology of Formations Used for Deep- Well Injection. Texas Gulf Coast: University of Texas at Austin Bureau of Economic Geology, Austin, 204 p. Report Prepared for the U.S. Environmental Protection Agency under Cooperative Agreement No. CR812786-01-0.

22. Doornhof, D., Kristiansen, T.G., Nagel, N.B., Pattillo, P.D. and Sayers, C. (2006) Compaction and Subsidence. Oilfield Review, 18, 50-68.

23. Poland, J.F. and Davis, G.H. (1969) Land Subsidence Due to Withdrawal of Fluids. Reviews in Engineering Geology, 2, 187-269. http://dx.doi.org/10.1130/REG2-p187

24. Gorshkov, S.P. (1998) Conceptual Basics of Geoecology. Smolensk State University, Smolensk.

25. Adushkin, V.V., Rodionov, V.N. and Turuntaev, S.B. (2000) Seismicity of Hydrocarbons Fields. Oil Review, 5, 4-15.

26. Kononova, M. (2011) Earthquakes. Characteristic. Examples. St. Petersburg State University, St. Petersburg.

27. Nikolaev, A.V. (1995) Problems of Induced Seismicity. In: Nikolaev, A.V. and Galkin, I.N., Eds., Induced Seismicity, Nauka, Moscow, 5-15.

28. Yashenko, V.R. (1989) Geodesic Researches of Vertical Movements of the Earth Crust. Nedra, Moscow, 192 p.

29. Yashenko, V.R. and Yambayev, H.K. (2006) Geodesy and Eternal Mysteries of the Earth Crust Movements. Geoprofi, 4, 61-66.

30. Lilienberg, D.A., Guseinzadeh, O.D., Kuliyev, F.T., Shirinov, N.Sh. and Yashenko, V.R. (1980) Complex Studies of Present Day Tectonic Movements on the Geodynamic Polygons of Azerbaijan. In: Bulanzhe, Yu.D., Lilienberg, D.A. and Podstrigach, Ya.S., Eds., Present Day Movements of the Earth Crust. Theory, Methods, Forecast, Nauka, Moscow, 165-174.

31. Muzipov, H.N., Yerka, B.A. and Illarionova, E.G. (2009) Environmental Safety of Plants of the Exploration and Production Sector. Neftyanoe Khozaystvo, 1, 92-94.

Citations

CHAPTER 1

Duanxin Chen, Shiguo Wu, Xiujuan Wang, and Fuliang Lv, "Seismic Expression of Polygonal Faults and Its Impact on Fluid Flow Migration for Gas Hydrates Formation in Deep Water of the South China Sea," Journal of Geological Research, vol. 2011, Article ID 384785, 7 pages, 2011. doi:10.1155/2011/384785.

CHAPTER 2

Zoulin Liu and Stephen M. J. Moysey, "The Dependence of Electrical Resistivity-Saturation Relationships on Multiphase Flow Instability," ISRN Geophysics, vol. 2012, Article ID 270750, 10 pages, 2012. doi:10.5402/2012/270750.

CHAPTER 3

Antonio Fernando Menezes Freire, Ryo Matsumoto, and Fumio Akiba, "Geochemical Analysis as a Complementary Tool to Estimate the Uplift of Sediments Caused by Shallow Gas Hydrates in Mounds at the Seafloor of Joetsu Basin, Eastern Margin of the Japan Sea," Journal of Geological Research, vol. 2012, Article ID 839840, 14 pages, 2012. doi:10.1155/2012/839840.

CHAPTER 4

Othniel K. Likkason (2014). Exploring and Using the Magnetic Methods, Advanced Geoscience Remote Sensing, Prof. Maged Marghany (Ed.), ISBN: 978-953-51-1581-6, InTech, DOI: 10.5772/57163.

CHAPTER 5

Maged Marghany (2014). Simulation of Tsunami Impact on Sea Surface Salinity along Banda Aceh Coastal Waters, Indonesia, Advanced Geoscience Remote Sensing, Prof. Maged Marghany (Ed.), ISBN: 978-953-51-1581-6, In Tech, DOI: 10.5772/58570.

CHAPTER 6

S. J. Anderson (2014). HF Radar Network Design for Remote Sensing of the South China Sea, Advanced Geoscience Remote Sensing, Prof. Maged Marghany (Ed.), ISBN: 978-953-51-1581-6, InTech, DOI: 10.5772/57599.

CHAPTER 7

M. F. Fingas, "Studies on the Evaporation Regulation Mechanisms of Crude Oil and Petroleum Products," *Advances in Chemical Engineering and Science*, Vol. 2 No. 2, 2012, pp. 246-256. doi: 10.4236/aces.2012.22029.

CHAPTER 8

KIM, J. , MIN, K. AND YIM, E. (2014) EXPERIMENTAL INVESTIGATION OF FUEL QUALITY AND CONTAMINANT MATERIALS FROM LIQUEFIED PETROLEUM FUEL. *ENGINEERING*, **6**, 644-653. DOI: 10.4236/ENG.2014.610064.

CHAPTER 9

Feyzullayev, A. and Ibragimov, V. (2014) Environmental Consequences of Long-Term Development of Petroleum Fields, Absheron p-la, Azerbaijan, Case History.*Journal of Environmental Protection*, 5, 1603-1610. doi:10.4236/jep.2014.517151.

Index